SpringerBriefs in Applied Sciences and Technology

SpringerBriefs present concise summaries of cutting-edge research and practical applications across a wide spectrum of fields. Featuring compact volumes of 50 to 125 pages, the series covers a range of content from professional to academic.

Typical publications can be:

- A timely report of state-of-the art methods
- An introduction to or a manual for the application of mathematical or computer techniques
- A bridge between new research results, as published in journal articles
- A snapshot of a hot or emerging topic
- An in-depth case study
- A presentation of core concepts that students must understand in order to make independent contributions

SpringerBriefs are characterized by fast, global electronic dissemination, standard publishing contracts, standardized manuscript preparation and formatting guidelines, and expedited production schedules.

On the one hand, **SpringerBriefs in Applied Sciences and Technology** are devoted to the publication of fundamentals and applications within the different classical engineering disciplines as well as in interdisciplinary fields that recently emerged between these areas. On the other hand, as the boundary separating fundamental research and applied technology is more and more dissolving, this series is particularly open to trans-disciplinary topics between fundamental science and engineering.

Indexed by EI-Compendex, SCOPUS and Springerlink.

Christoph Hoppe

Commissioning Management in Offshore Wind Projects

The Ultimate Guide

Springer

Christoph Hoppe
Hude, Germany

ISSN 2191-530X ISSN 2191-5318 (electronic)
SpringerBriefs in Applied Sciences and Technology
ISBN 978-3-032-05357-2 ISBN 978-3-032-05358-9 (eBook)
https://doi.org/10.1007/978-3-032-05358-9

This Springer imprint is published by the registered company Springer Nature Switzerland AG
The registered company address is: Gewerbestrasse 11, 6330 Cham, Switzerland

Preface

To the Readers

My name is Christoph Hoppe, and I have been working in the offshore wind industry for over 10 years. Primarily, I have served as a Commissioning Manager and as a Responsible Electrical Specialist—General Senior Authorized Person (G-SAP). Together with my clients and employers, I have realized offshore wind projects around the world and was among the pioneers permitted to construct the first commercial offshore wind farm in Taiwan/Asia. It was an exciting time full of new experiences. With this book, I aim to support all those who are still gaining their experience in this field. I want to emphasize that this book neither claims to be exhaustive nor the only true path to Commissioning Management. However, based on my experiences and successes, these contents can take you a significant step further and should enable you to complete a project with a certain degree of success.

Commissioning Management must always be adapted and can change significantly from phase to phase within a project. So, I can only clearly state that this book will not be a detailed guide on how you should manage your commissioning team. However, I am convinced that it will serve as an inspiration for you, from which you can draw one thing or another and implement it in your projects.

Special Thanks

I would like to extend special thanks to two people: Peter Krieglmeier and Burkhard Maass. These two outstanding Project Managers have accompanied, supported, and shaped me over 10 years in various projects. Without them, I wouldn't be where I am today. Peter and Burkhard, I thank you. You have been not only colleagues, employees, and teammates but have also become friends—friends I could always rely on. There was no one else with whom I would have preferred to realize projects.

Hude, Germany Christoph Hoppe

Competing Interests The author has no competing interests to declare that are relevant to the content of this manuscript.

Contents

Abbreviations

Below is a list of abbreviations used in the English version of the book, along with their full forms:

3DLAR	3-Day Look-Ahead Report
AI	Artificial Intelligence
AP	Authorized Person
AR	Augmented Reality
CTV	Crew Transfer Vessel
DNV	Det Norske Veritas
DPR	Daily Progress Report
HSE	Health, Safety, and Environment
IEC	International Electrotechnical Commission
ISO	International Organization for Standardization
MCC	Marine Coordination Center
MoM	Minutes of Meeting
O&M	Operations and Maintenance
OCC	Operation Control Center
P50	50th Percentile Scenario
P90	90th Percentile Scenario
PPE	Personal Protective Equipment
SAP	Senior Authorized Person
SOV	Service Operation Vessel
SST	Self-Sustainable Turbine
TSO	Transmission System Operator
VR	Virtual Reality
WPR	Weekly Progress Report
WTG	Wind Turbine Generator

List of Figures

Part I
Fundamentals of Commissioning Management

Commissioning Management forms the heart of the commissioning processes of an offshore wind project. It bridges the construction phase with operations and ensures that all systems and components of a wind farm function reliably, safely, and efficiently. In this section, we delve into the central fundamentals of Commissioning Management—from basic principles and roles and responsibilities to strategic approaches. Part I provides a comprehensive understanding of the structural and technical requirements necessary to successfully and sustainably commission offshore wind projects.

Chapter 1
Introduction to Offshore Wind Projects

1.1 Overview of the Offshore Wind Industry

The offshore wind industry has evolved over the past decades from its humble beginnings into a central pillar of the global energy transition. This section provides a comprehensive overview of the development, current status, and future prospects of offshore wind energy.

1.1.1 Historical Overview

The history of offshore wind energy began in 1991 with the commissioning of the world's first offshore wind farm off the coast of Vindeby in Denmark. This pioneering park consisted of eleven turbines with a total capacity of 5 MW and marked the beginning of a new era in energy generation [1].

In the following years, the offshore wind industry expanded continuously. In 2002, the Horns Rev 1 wind farm with a capacity of 160 MW was constructed in Denmark—the largest of its kind at the time. Germany entered the offshore wind energy sector in 2009 with the test field "alpha ventus," which served as a pilot project for future developments [2].

1.1.2 Current State of the Offshore Wind Industry

In the last five years, offshore wind energy has experienced impressive growth worldwide. By the end of 2023, the global installed capacity reached approximately 67 GW, spread across 19 countries. In 2023 alone, 25 new wind farms were connected to the grid, and another three dozen were under construction.

C. Hoppe, *Commissioning Management in Offshore Wind Projects*,
SpringerBriefs in Applied Sciences and Technology,
https://doi.org/10.1007/978-3-032-05358-9_1

1.1.3 Europe

Europe remains a leading player in offshore wind energy. The United Kingdom boasts the largest installed capacity with over 10 GW, followed by Germany with about 8.5 GW. Denmark, the Netherlands, and Belgium also contribute significantly to Europe's offshore wind capacity [3].

In 2023, construction began on the German offshore wind farms Riffgrund 3 and Gode Wind 3 in the North Sea. For the first time, mega wind turbines with a capacity of 11 MW each from Siemens Gamesa are being used [4].

1.1.4 Asia

China has evolved into the world's largest offshore wind market in recent years. By the end of 2023, the installed capacity reached over 20 GW, accounting for about one-third of the global capacity. The Chinese government supports expansion through strong political measures and low financing costs [5].

Japan and South Korea also have ambitious plans for expanding offshore wind energy, with several projects in the planning and construction phases.

1.1.5 USA

The USA has intensified efforts in recent years to enter the offshore wind sector. Although the installed capacity is still relatively low, several large projects are planned, especially on the East Coast. The Biden administration has set ambitious goals for the expansion of offshore wind energy to achieve the country's climate targets [5].

1.1.6 Technological Developments

The offshore wind industry has made significant technological advances in recent years. The development of floating offshore wind farms enables the use of deeper waters that previously remained untapped. For example, France plans to expand offshore wind energy and aims to achieve a total of 40 GW of offshore wind power capacity by 2050 [4].

The size and efficiency of wind turbines have also increased. Modern turbines achieve capacities of over 10 MW, significantly boosting power generation per unit. For instance, Vestas turbines with 9.5 MW each were installed in the German offshore wind farm Arcadis Ost 1 in 2023 [4].

1.1.7 Challenges and Perspectives

Despite the impressive growth, the offshore wind industry faces several challenges. Rising costs, project delays, and limited investments in the supply chain could lead to ambitious goals of some countries not being met. For example, it is expected that the USA and European countries might miss their targets for 2030 [5].

Nevertheless, offshore wind energy remains a central component of global efforts to reduce CO_2 emissions and combat climate change. With continuous investments, technological innovations, and supportive political frameworks, the industry is expected to continue growing and make a significant contribution to a sustainable energy future.

1.2 Lifecycle of an Offshore Wind Project

The lifecycle of an offshore wind project is a multifaceted and complex process organized into clearly defined phases—from initial planning to decommissioning. Each of these phases presents specific requirements for technical expertise, coordination, and strategic decision-making. Careful execution of each phase is crucial to secure the long-term success of such a project.

1.2.1 Planning and Development

The foundation of any offshore wind project is laid in the planning and development phase. Everything begins with the selection of a suitable site, which must offer optimal wind resources while considering aspects like water depth, distance to shore, and environmental compatibility. Developers often face the challenge of minimizing potential conflicts with other uses such as shipping or fishing [6].

Additionally, feasibility studies are conducted to analyze the technical, economic, and environmental viability of the project. These studies include wind measurements, geotechnical investigations of the seabed, and extensive environmental impact assessments [7]

In parallel, the often complex permitting process begins. Developers must meet numerous regulatory requirements to obtain the necessary construction permits and offshore licenses. This phase is also marked by securing financial resources. Project developers frequently resort to public–private partnerships or other innovative financing models to manage the high capital commitment of such large scale projects [7].

The combination of solid planning and secured financing creates the basis for successful implementation.

1.2.2 Design and Engineering

After the successful development phase, the technical foundation of the project is established in the design and engineering phase. The choice of wind turbines, foundations, and electrical components is tailored to the site conditions [8].

Simultaneously, grid connection solutions must be developed to safely transmit the generated electricity from the offshore facility to land. This includes planning offshore substations, export cables, and onshore infrastructures [7].

Detailed project planning plays a crucial role here. Schedules, logistics concepts, and resource allocations are designed so that all activities are optimally coordinated. Every element—from the construction time of the foundations to the installation of the turbines—must be precisely planned to avoid delays and cost increases [6].

1.2.3 Procurement and Manufacturing

With the completion of the planning work, the procurement and manufacturing phase begins. Suppliers are selected, and the production of necessary components commences. Tendering processes play a central role in ensuring high-quality and economical offerings [9].

At the same time, the manufacturing of turbines, foundations, and cables starts. Manufacturers work closely with developers to ensure that all components meet strict specifications and timelines. This phase is a logistical masterpiece, as timely delivery to offshore sites critically impacts the project schedule [10].

1.2.4 Construction and Installation

The actual construction of the wind farm begins with the installation of the foundations. Depending on site conditions, different foundation types like monopiles, jackets, or gravity foundations are used [11].

Simultaneously, the internal array cabling, which connects the turbines with each other, and the export cables, which transport the generated electricity to land, are laid [12].

These tasks require the use of highly specialized installation vessels and experienced offshore personnel.

Once the foundations and cables are completed, the installation of the turbines begins. Each turbine is carefully mounted on its foundation, with modern units reaching heights of over 200 m. After assembly, commissioning takes place, during which all systems are tested and released for operation. This phase is crucial to ensure that the wind farm operates efficiently and safely [11].

1.2.5 Operation and Maintenance

After successful commissioning, the project enters the operational phase, which typically lasts 20–25 years. During this time, the wind farm is continuously monitored to maximize energy production and detect potential problems early [6].

State-of-the-art SCADA systems (Supervisory Control and Data Acquisition) enable real-time monitoring and control of the facilities. Maintenance work, both scheduled and unscheduled, is essential to ensure the longevity of the assets. Operators increasingly rely on data analysis and advanced algorithms to further enhance turbine efficiency [13].

1.2.6 Decommissioning and Dismantling

At the end of their lifespan, wind farms are decommissioned. This phase begins with developing a detailed decommissioning plan that includes the removal of facilities and the restoration of the seabed [9].

The dismantling is carried out in compliance with strict environmental and safety standards. The recyclability of materials, especially the turbine blades, is gaining increasing importance as innovative processes are developed to improve the sustainability of decommissioning. This final phase marks the end of the lifecycle of an offshore wind project but also opens the door for new projects at the same sites [14].

1.3 Role of Commissioning Management

1.3.1 Introduction to the Role of Commissioning Management

Commissioning Management is a central component of any offshore wind project and plays a crucial role in ensuring that all systems and components of a wind farm are properly installed, tested, and ready for operation before commercial activities commence. Through precise planning, effective coordination, and close collaboration with all project participants, Commissioning Management ensures compliance with quality standards, safety regulations, and schedules. It thus contributes significantly to the efficiency, safety, and long-term reliability of an offshore wind farm.

1.3.2 *Planning and Preparation*

In the initial phase, Commissioning Management undertakes detailed planning and preparation. This includes creating a comprehensive commissioning plan that defines all tests, inspections, and milestones. Alongside a clearly structured schedule, resources such as personnel, equipment, and technologies are optimally allocated. It is particularly important to precisely define responsibilities within the team to avoid overlaps and maximize process efficiency.

1.3.3 *Coordination of Commissioning*

During the implementation phase, Commissioning Management coordinates the commissioning activities in close collaboration with construction managers, engineers, suppliers, and other stakeholders. This coordination encompasses monitoring on-site progress and adjusting plans if unforeseen challenges arise. A central component is aligning different trades to ensure that all systems work seamlessly together and meet specified requirements. Reviewing construction progress and quality standards plays a decisive role here.

1.3.4 *Conducting Tests and Inspections*

Another essential task of Commissioning Management is conducting and supervising tests and inspections. Mechanical, electrical, and functional tests of all components—including turbines, foundations, cables, and substations—are organized. These tests aim to verify the proper function, safety, and efficiency of the systems. Ensuring that each component meets project-specific requirements and that any deviations are detected and corrected early is particularly important.

1.3.5 *Safety Management*

In addition to technical supervision, Commissioning Management plays a key role in safety management. Throughout the commissioning phase, a strong focus is placed on carrying out all activities in strict compliance with safety regulations. This includes identifying potential risks, implementing safety protocols, and continuous training of personnel. Especially in offshore environments—with specific hazards like high voltage, extreme weather conditions, and limited access—rigorous safety management is essential.

1.3.6 Documentation and Reporting

Another important aspect of Commissioning Management is documentation and reporting. Throughout the entire process, comprehensive documentation is maintained, covering progress, test results, and any deviations. These reports are crucial not only for internal tracking and process optimization but also serve as evidence of compliance with legal requirements, quality standards, and insurance conditions. Structured and detailed documentation also facilitates the handover of the project to the operations team.

1.3.7 Handover to Operations

The handover to operations is another central responsibility within Commissioning Management. After completing all tests and inspections, the team ensures that the wind farm is smoothly transferred to the operations department. This process includes providing all relevant documentation, training operational personnel, and ensuring that all parties are familiar with the installed systems and their operation. The handover marks the official completion of the commissioning phase and ensures that the wind farm can commence commercial operation efficiently and without interruptions.

1.3.8 Importance of Commissioning Management

Overall, Commissioning Management acts as a connecting element between the construction and operation of an offshore wind farm. Through precise planning, coordinated implementation, and continuous monitoring of all steps, it significantly contributes to enabling offshore wind projects to reach their full potential and make a substantial contribution to the energy transition. This role requires a high level of technical expertise, organizational skills, and teamwork to successfully overcome the various challenges of an offshore project.

References

1. Oersted, *30 Jahre Offshore-Windenergie: Von Dänemark in die ganze Welt* (2021). Von https://orsted.de/presse-media/news/2021/09/30-jahre-offshore-wind abgerufen
2. Internationales Wirtschaftsforum Regenerative Energien (IWR)/IWR.de GmbH, *Chronologie Offshore Windenergie* (2017). Von https://www.offshore-windindustrie.de/news/chronologie abgerufen
3. Oersted, *Diese Länder setzen auf Offshore-Wind* (2022). Von https://energiewinde.orsted.de/energiewirtschaft/offshore-windparks-liste-laender-weltweit abgerufen

4. Internationales Wirtschaftsforum Regenerative Energien (IWR)/IWR.de GmbH, *Offshore Windparks in Deutschland* (2024). Von https://www.offshore-windindustrie.de/windparks/deutschland abgerufen
5. N.G. Chestney, *Global Offshore Wind Industry Poised to Miss Big Targets as Obstacles Mount* (2024). Von https://www.reuters.com/sustainability/global-offshore-wind-industry-poised-miss-big-targets-obstacles-mount-2024-11-18/ abgerufen
6. U.S. Department of Energy's Wind Energy Technologies Office, *Wind Project Site Selection* (2024). Von https://windexchange.energy.gov/projects/site-selection abgerufen
7. WindEurope asbl/vzw, *Offshore Wind Energy* (2024). Von https://windeurope.org/policy/topics/offshore-wind-energy/ abgerufen
8. GWEC—Global Wind Energy Council, *Global Offshore Wind Report 2024* (2024). Von https://www.gwec.net/reports/globalofffshorewindreport abgerufen
9. J. Dhavle, *Floating Offshore Wind Outlook* (International Renewable Energy Agency, Abu Dhabi, 2024). Von https://www.irena.org/Publications/2024/Jul/Floating-offshore-wind-outlook abgerufen
10. Siemens Gamesa Renewable Energy, S.A.U, *Commissioning and Troubleshooting Self Sustained Turbine Training* (2024). Von https://training.siemensgamesa.com/sginternet/beschreibung1.jsp?Course=CA0WI70 abgerufen
11. Det Norske Veritas, *Offshore Wind* (2024). Von https://www.dnv.com/power-renewables/themes/offshore-wind/ abgerufen
12. Prysmian S.p.A., *Power Grid* (2024). Von https://australia.prysmian.com/markets/power-grid abgerufen
13. Vestas Wind Systems A/S, *Offshore Wind Turbines* (2024). Von https://www.vestas.com/en/energy-solutions/offshore-wind-turbines abgerufen
14. General Electric Company, *Second Wind: GE and Partners Seek to Recycle Turbine Blades in Europe* (2021). Von https://www.ge.com/news/reports/second-wind-ge-and-partners-seek-to-recycle-turbine-blades-in-europe-0 abgerufen

Chapter 2
Organization and Roles

The successful execution of an offshore wind project significantly depends on a clear organizational structure and the definition of specific roles. Given the immense complexity of these projects—which bring numerous technical, logistical, and safety-related challenges from the construction phase to the operational phase—it's essential to assign team members' responsibilities precisely. This is especially true for Commissioning Management, which serves as the crucial link between the construction phase and the subsequent operation of a wind farm. Establishing clear reporting lines and efficient escalation mechanisms is not only an organizational necessity but also a key to adhering to schedules, budgets, and the highest safety standards.

2.1 Key Stakeholders

In an offshore wind project, many different stakeholders collaborate, each bringing their own responsibilities and perspectives. Developers, often consisting of project companies or consortia, undertake the planning and financing of the project. This role requires careful coordination with investors and ensuring that regulatory requirements are met. Another central player is the so-called EPC (Engineering, Procurement, Construction), who, as the main contractor, is responsible for technical planning, procurement, and construction. EPCs often coordinate multiple subcontractors responsible for specific tasks like cable laying, substation construction, or turbine installation.

Furthermore, transmission system operators, also known as TSOs, play a crucial role as they realize the connection of the wind farm to the power grid. Their task is to design the infrastructure so that the generated electricity can be reliably fed into the grid. Regulatory authorities form another important component of the stakeholder network, as they monitor compliance with legal requirements and ensure that operations are both environmentally and safety-wise acceptable. This close interweaving

C. Hoppe, *Commissioning Management in Offshore Wind Projects*,
SpringerBriefs in Applied Sciences and Technology,
https://doi.org/10.1007/978-3-032-05358-9_2

of the various interest groups illustrates how important smooth coordination and clear communication paths are.

2.2 Roles and Responsibilities

Within Commissioning Management, which forms the interface between construction and operation, clearly defined roles assume responsibility for planning, executing, and monitoring the commissioning processes. The **General Senior Authorized Person** (GSAP) stands at the top of this team and takes overall responsibility for all electrical work within the project. This role is crucial to ensure safety and compliance with operational regulations throughout the entire commissioning phase. The GSAP also carries the responsibility for the approval of switching operations and ensures that risk assessments are properly conducted. In case of safety-relevant questions or deviations from planned procedures, the GSAP is the primary escalation point and coordinates with higher management or the HSE teams (Health, Safety, Environment).

The **Senior Authorized Person** (SAP) is positioned directly under the GSAP and takes on operational responsibility for the practical execution of high and medium voltage switching operations. This role requires not only technical expertise but also close coordination with other team members. The SAP creates detailed switching plans to ensure that all electrical work is carried out in a controlled and risk-free manner. At the same time, they monitor compliance with Lockout-Tagout (LOTO) processes, which ensure that electrical installations are safely isolated during work. In addition to the practical execution of switching operations, the SAP also documents all relevant information and prepares reports that are significant for both internal purposes and official inspections.

The **Authorized Person** (AP) supports the SAP in implementing specific tasks. This role focuses on the practical execution of preparatory work, such as inspecting installations and setting up the work environment for tests. The AP ensures that all necessary conditions for carrying out the work are met and reports potential safety risks or deviations from standards to the SAP or GSAP.

The **Competent Person** (CP) is another important role within Commissioning Management. They perform specialized activities such as installing specific components or assisting with complex tests. The CP works under the guidance of the SAP or AP and ensures that all work is carried out safely and efficiently. Another central responsibility of the CP is the maintenance and calibration of test equipment to ensure that results are reliable.

The **Operation Control Center Manager** (OCC Manager) oversees all activities during commissioning from a central control point. This role serves as a link between offshore teams and onshore management and coordinates switching operations, tests, and inspections. The OCC Manager documents every activity and ensures a seamless flow of information to identify and resolve delays or safety risks early.

2.2.1 Other Responsible Roles in Offshore Wind Projects

Besides the roles within Commissioning Management, there are other positions essential for the overall success of an offshore wind project. These include the **HSE Manager**, who is responsible for monitoring compliance with safety and environmental standards, and the **Risk and Interface Manager**, who identifies potential risks and coordinates collaboration between various stakeholders. The **Logistics Manager** plays a crucial role in planning and implementing offshore logistics, while the **Quality Manager** ensures that all work meets the established quality standards. The **Marine Coordinator** is responsible for coordinating vessel traffic and safety during offshore work. Although these roles are only briefly mentioned in this chapter and not considered in detail, they, along with other responsible roles like the Construction Manager, are integral components of a successful offshore wind project.

2.3 Organizational Structure of a Commissioning Team

The organizational structure of Commissioning Management is hierarchically built to enable clear assignment of responsibilities and reporting lines. At the top is the GSAP, who is responsible for the strategic leadership of the team. The SAPs take over operational control and coordinate the work of the APs and CPs. The OCC Manager ensures that all activities are centrally monitored and effectively communicated. The organizational structure also integrates clear escalation paths to ensure that problems or safety incidents are promptly reported and addressed. A well-defined structure is essential to coordinate the numerous parallel activities within commissioning and to ensure a smooth process.

The role of Commissioning Management and its embedding within a clear organizational structure are essential prerequisites for the success of an offshore wind project. By precisely delineating responsibilities and integrating specialized roles, not only is efficiency increased, but the safety and quality of the work are also ensured. Collaboration with other key roles in the project, such as HSE or Risk Managers, strengthens the overall structure and contributes to effectively overcoming challenges. This makes Commissioning Management an indispensable building block for every successful offshore wind project.

Part II
Processes and Tools

Efficient processes and the use of modern tools are essential for successful Commissioning Management. Part II sheds light on the structured planning and execution of commissioning activities, as well as the technologies that support these processes. From creating detailed schedules to implementing digital tools, it describes the crucial mechanisms and methods that ensure transparency, safety, and quality in offshore wind projects. This section demonstrates how processes and tools seamlessly work together to manage the complexity of modern projects and secure their long-term success.

Chapter 3
Planning and Preparation

Planning and preparation is the fundamental phase of any offshore wind project. In this phase, the framework for all subsequent activities is established. Without precise planning, delays, quality issues, or safety risks could jeopardize the entire project. A solid foundation begins with the development of a comprehensive **Commissioning Management Strategy**, which serves as the strategic guiding document and forms the basis for all further planning. Building on this, the **Commissioning Management Plan** is developed, defining operational details and processes. This is complemented by a series of **Commissioning Management Documents** that cover specific aspects of commissioning and provide clear guidance for the entire team.

3.1 The Commissioning Management Strategy

The Commissioning Management Strategy is the overarching document that sets out the strategic direction and overall goals of Commissioning Management. It acts as a "Project Charter" for commissioning and defines the organizational structure as well as the guidelines for execution. A central aim of the Commissioning Management Strategy is to ensure that all project participants—from internal teams to external subcontractors—pursue a unified vision and work towards common goals.

The Commissioning Management Strategy begins with a clear description of the project objectives, including adherence to schedules, ensuring high safety standards, meeting technical requirements, and minimizing risks. Furthermore, it defines the main roles in the project, reporting lines, and escalation mechanisms to optimize collaboration and quickly resolve potential conflicts. It also contains an analysis of the key risks, such as weather-related delays, potential material shortages, or interface problems between different trades.

A particular focus lies on defining the communication strategy. The Commissioning Management Strategy stipulates how information is exchanged between

C. Hoppe, *Commissioning Management in Offshore Wind Projects*,
SpringerBriefs in Applied Sciences and Technology,
https://doi.org/10.1007/978-3-032-05358-9_3

offshore and onshore teams, as well as with external stakeholders like authorities and grid operators. Regular status reports, structured meetings, and standardized communication protocols are essential elements. The Commissioning Management Strategy is regularly reviewed and updated throughout the project to flexibly respond to changes.

3.2 The Commissioning Management Plan

Building on the Commissioning Management Strategy, the Commissioning Management Plan is developed, serving as the operational guide for commissioning. While the Strategy provides the strategic framework, the Commissioning Management Plan focuses on the detailed planning and execution of individual processes. It defines concrete schedules, resource requirements, and specific milestones that must be achieved to enable the handover of the wind farm to the operational phase.

A central element of the Commissioning Management Plan is the description of responsibilities. The respective roles within the team—from the General Senior Authorized Person to the Competent Person—are clearly delineated to avoid overlaps and ensure efficiency. Additionally, the Commissioning Management Plan specifies how resources such as personnel, equipment, and logistics are utilized. This includes planning offshore transports, ensuring the availability of specialized testing equipment, and making certain that all team members have the required training.

The Commissioning Management Plan also includes a detailed schedule covering all commissioning activities from the start of testing to the gradual energization. This schedule identifies critical milestones, such as the execution of Factory Acceptance Tests (FAT), Site Acceptance Tests (SAT), and Harbour Acceptance Tests (HAT). By identifying dependencies between different phases, potential delays can be recognized early and minimized.

3.3 Commissioning Management Documents

The Commissioning Management Documents complement the Commissioning Management Plan and cover specific aspects of commissioning. They serve as practical tools to ensure compliance with planned processes and guarantee consistent execution.

3.3.1 Schedule and Progress Monitoring

A comprehensive schedule is one of the central Commissioning Management Documents. It provides a detailed overview of all planned activities and defines their chronological sequence as well as the respective responsibilities. This schedule is continuously updated to monitor progress and ensure that critical milestones, such as the completion of cable installations or the commissioning of offshore substations, are achieved on time. Progress reports based on this schedule regularly inform all project participants about the current status and possible delays.

3.3.2 *Safety Procedures*

Safety in offshore wind projects has the highest priority, especially during the commissioning phase when teams work under demanding conditions. Safety procedures are an essential part of the Commissioning Management Documents and define the processes and guidelines required to prevent accidents and ensure the safe operation of the facilities. These procedures cover a broad spectrum, from general safety measures to specific requirements for high-voltage switching operations and electrical work.

A central element of the safety procedures is the Lockout-Tagout (LOTO) processes, which ensure that no energy supply reaches the affected systems during maintenance and installation work. These procedures involve the physical isolation of energy sources, the placement of warning signs, and the keeping of logs to ensure that all involved personnel are informed about the status of the systems. LOTO is an indispensable component of any safety strategy and is strictly enforced.

Furthermore, **Method Statements** for high-voltage switching operations play a crucial role. These documents describe in detail the steps to be followed when performing switching operations on high-voltage installations. Method Statements encompass the planning, execution, and verification of switching operations, as well as measures to minimize risks. They define clear responsibilities and contain specific instructions to ensure that all work is carried out safely and efficiently. These Method Statements are created in close coordination with the Senior Authorized Person, who is responsible for supervising and approving the switching operations.

Another important aspect of the safety procedures is the **Electrical Safety Procedures**, specifically developed for the requirements of electrical work. These procedures include regulations for the safe handling of high and medium-voltage systems, the use of personal protective equipment (PPE), and the execution of tests on electrical components. For example, insulation tests are conducted on each electrical installation to ensure that no dangerous voltages are present before work begins. Electrical Safety Procedures also set standards for the qualification of involved personnel and prescribe regular training to ensure compliance with current safety regulations.

In close collaboration with the HSE (Health, Safety, Environment) department, emergency plans are integrated into the safety procedures to ensure that unforeseen events can be responded to quickly and effectively. These plans include evacuation procedures for offshore structures, first-aid measures, and coordination with rescue services. Offshore wind projects are often affected by extreme weather conditions, so emergency plans also cover scenarios such as sudden weather changes or evacuations due to storms.

The safety procedures also encompass comprehensive training programs to ensure that all team members are familiar with the specific requirements and risks of their tasks. These trainings include theoretical and practical units, such as conducting rescue exercises in high-voltage environments or working safely in confined spaces. The goal is to prepare employees for potential hazards and enable them to act safely even in critical situations.

The implementation of these procedures is supported by regular audits and inspections. These reviews serve to ensure that all safety guidelines are being followed and allow for the identification of areas for improvement. Experiences from previous projects and lessons learned flow into the continuous optimization of safety procedures to further minimize future risks.

In summary, the safety procedures provide a comprehensive foundation to ensure safety during the commissioning phase. They cover all relevant areas—from planning to execution to training and emergency preparedness. By strictly adhering to these procedures, not only is the protection of the teams ensured, but the foundation for the safe and successful operation of the offshore wind farm is also laid.

3.4 Test Procedures and Checklists

Test procedures and associated checklists form the backbone of quality assurance in Commissioning Management. For each component of the wind farm—be it a turbine, a cable, or a substation—specific test methods are defined to ensure that the respective systems function properly. The checklists serve as documentation tools to record the results of the tests and ensure consistent execution. They also provide clear guidance for technicians on-site and facilitate the tracking of issues.

3.4.1 Non-conformity Reports

A **Non-conformity Report** (NCR) is an indispensable document in Commissioning Management, used to systematically record, analyze, and rectify deviations from established specifications, standards, or processes. NCRs are created when defects or deviations are identified during the commissioning phase that could impair the safety, quality, or performance of the facility.

The NCR describes the deviation in detail, including the affected component or system, the nature of the deviation, and the circumstances under which it was identified. The report is supplemented by evidence such as photos, test protocols, or measurement data. After documentation, an assessment of the impacts on the project schedule, operational safety, and long-term reliability is conducted.

Commissioning Management uses NCRs not only for problem-solving but also as a tool for continuous improvement. They make it possible to identify patterns of recurring problems and initiate appropriate measures to prevent future deviations. Thus, NCRs ensure that the commissioning processes are completed efficiently and in accordance with the highest standards.

3.4.2 Resource and Training Plans

The resource and training plans ensure that all team members have the necessary qualifications and tools to effectively fulfill their tasks. The training plans encompass both technical training, such as handling test equipment, and safety-related training, for example in rescue techniques in offshore environments. The resource plans specify the required equipment—from protective gear to specialized tools—and ensure their availability.

3.4.3 Communication Matrix

Effective communication is a central success factor for Commissioning Management. An agreed-upon communication matrix defines how information is exchanged between different teams and stakeholders. It outlines regular meetings, reporting formats, and escalation mechanisms to ensure that all parties are always up to date. This matrix promotes smooth collaboration and prevents misunderstandings that could lead to delays.

3.4.4 Schedules and Dependencies

A structured and precise schedule is an integral part of planning in Commissioning Management. It serves as an operational guideline for all participants and ensures that critical milestones are reached on time. The creation and maintenance of the schedule are carried out in close collaboration with the **Schedule Manager**, whose expertise in scheduling and resource coordination plays a key role. The schedule, often visualized as a **Project Schedule** in the form of a **Gantt chart**, provides a clear overview of the chronological sequence of all activities, their dependencies, and the current progress.

The Gantt chart is an indispensable tool in Commissioning Management. It allows the graphical representation of the entire project course—from the initial tests to the final energization of the wind farm. Each bar in the chart represents a specific task or phase, such as conducting Factory Acceptance Tests (FAT), installing offshore cables, or performing Site Acceptance Tests (SAT). By depicting dependencies—such as arrows connecting activities—it becomes clear which tasks depend on each other and which activities can run in parallel. This helps the commissioning team identify critical paths and ensure that delays in one task do not have cascading effects on other phases.

Collaboration between the Commissioning Manager and the Schedule Manager is crucial to efficiently coordinate dependencies between different trades and phases. A good example of this is the installation and testing of export cables, which must be completed before the substations can be commissioned. The Schedule Manager analyzes these dependencies and integrates them into the schedule to recognize and avoid potential conflicts early. At the same time, the Schedule Manager ensures that changes in the schedule, such as those due to weather-related delays, are communicated promptly to the commissioning team so that alternative scenarios can be planned.

Another central aspect of scheduling is the application of scenario analyses, particularly the **P50** and **P90** approaches. These statistical models help create realistic schedules and assess the risks of delays. A P50 scenario describes a situation where there is a 50% probability of meeting the planned date. It represents the "median" expectation and is often used as a basis for operational planning. The P90 scenario, on the other hand, takes a more conservative approach and shows how the schedule develops under less optimal conditions. Here, the probability of exceeding the date is only 10%. The P90 scenario is often used for risk assessments and communication with stakeholders, as it offers a high degree of certainty.

These scenario analyses are particularly valuable when planning weather-dependent activities, such as transporting components to offshore sites or performing switching operations. For example, a P50 schedule for offshore transports during the summer period might be optimistic if the weather is stable. However, a P90 schedule also accounts for possible disruptions due to unforeseen weather changes and thus provides a reliable basis for risk planning.

In addition to scenario analyses, the schedule allows for detailed progress monitoring. The Commissioning Manager works closely with the Schedule Manager to regularly incorporate updates into the Gantt chart. These progress reports show which milestones have been achieved and help identify any deviations from the plan. If, for instance, a critical test is not conducted on time, early detection allows for corresponding countermeasures to be taken, such as rescheduling resources or prioritizing certain tasks.

The integration of a clearly structured and dynamic schedule creates transparency and ensures that all teams work in sync. By closely collaborating with the Schedule Manager, not only is efficiency increased, but also the foundation for a successful and timely commissioning is laid. A well-maintained Gantt chart, complemented by well-founded scenario analyses like P50 and P90, provides the necessary tools to master the complex dependencies in an offshore wind project and ensure long-term project success.

Chapter 4
Safety and Risk Management

Safety and risk management in offshore wind projects are essential components of successful project execution. They ensure not only the protection of the personnel involved but also compliance with technical standards and the long-term reliability of the installations. This chapter provides a detailed description of the collaboration between the HSE Manager and the Commissioning Manager, as well as key safety processes like Lockout-Tagout (LOTO), Validation of Clearance (VoC), and Sanction for Testing (SFT). Switching operations, particularly in high voltage, are addressed in a separate chapter to meet the high demands of this topic.

4.1 Roles and Collaboration: HSE Manager and Commissioning Manager

The HSE Manager bears the main responsibility for developing and implementing all safety-related processes in the project. This includes creating safety policies, conducting risk analyses, and monitoring compliance with regulations. Safety risks, which are mainly considered in this chapter, should not be confused with economic risks. The Risk Manager is responsible for identifying and assessing project-related economic risks. Their main task is to recognize potential cost increases early, establish extensive risk budgets, and develop appropriate countermeasures.

The Commissioning Manager works closely with both roles. In the field of electrical safety, they support the HSE Manager in developing and implementing specific safety measures. At the same time, they provide the HSE Manager with technical information and practical experience that feed into the risk analyses. This collaboration is essential to ensure integrated safety and risk management.

C. Hoppe, *Commissioning Management in Offshore Wind Projects*,
SpringerBriefs in Applied Sciences and Technology,
https://doi.org/10.1007/978-3-032-05358-9_4

4.2 Safety Procedures and Method Statements

4.2.1 Lockout-Tagout (LOTO): Safety Through Isolation

The LOTO procedure is a fundamental tool for controlling hazardous energy sources. Its goal is to ensure that no unauthorized or accidental energy supply occurs during work on electrical systems.

4.2.2 Practical Application in the Offshore Context

On offshore structures, the LOTO procedure is applied to all work on or with electrical systems. Isolating, for example, a transformer cell involves not only disconnecting the main circuits but also securing all control lines and discharging stored energies, such as the spring tension of circuit breakers. The SAP oversees this process and ensures that all isolation measures comply with applicable safety standards.

4.2.3 Detailed Procedure of the LOTO Process

1. **Preparation**: The LOTO responsible person, often an Authorized Person (AP) or Competent Person (CP), reviews all relevant circuit diagrams to ensure that all energy sources are identified. This includes hidden back-feeds or capacitive residual charges.
2. **Notification**: All affected parties—including the offshore team, the onshore control center, and the HSE Manager—are informed. Clear communication prevents misunderstandings and ensures coordinated execution.
3. **Shutdown**: High-voltage lines are switched off by the authorized switching technician before low-voltage components are deactivated. This is done according to detailed Method Statements that specify all necessary steps.
4. **Isolation**: Energy sources are physically disconnected and grounded by the CP or AP to exclude any back-feeding. The SAP conducts an inspection to confirm proper execution.
5. **Lockout and Tagging Devices**: Physical locks are applied and equipped with clearly labeled warning signs containing information about the responsible person and the isolation status.
6. **Energy Discharge**: Capacitors and other energy storage devices are completely discharged. Technicians use specialized discharge equipment to ensure no residual voltage remains.
7. **Verification**: The complete absence of voltage is verified using appropriate measuring devices. The SAP documents the results and releases the work area.

4.2.4 *Validation of Clearance (VoC): Ensuring a Hazard-Free Work Area*

Validation of Clearance is a central safety instrument that clearly defines the status of a work area. It confirms that all energy sources have been isolated and no electrical hazards exist. Sometimes, especially when working in the UK, the term **COI** (Certificate of Isolation) is used. Essentially, a COI and a VoC are the same.

4.2.5 *Reasons and Benefits*

The VoC minimizes the risk of accidents by ensuring the work area is free of voltage. Additionally, it serves as an official document communicating the current safety status and clearly defining responsibilities. This creates transparency and reduces the risk of communication errors.

4.2.6 *Procedure of the VoC*

After completing the LOTO measures, the absence of voltage is checked by the Senior Authorized Person. All measurement results and inspections are documented in an isolation certificate, which is released by the HSE Manager. Only after issuing the VoC may work begin.

4.2.7 *Sanction for Testing (SFT): Approval for Specific Tests*

Sanction for Testing is a formal process that approves the execution of specific tests. It serves to set clear boundaries for the tests and ensure that all involved personnel are informed about the planned activities.

The SFT describes the specific tests to be conducted and the safety measures to be observed. It defines test boundaries, voltage requirements, and the responsibilities of the involved technicians. A central component of the SFT is communication between the Commissioning Manager, the test leader, and the HSE Manager to ensure all safety precautions have been taken.

Chapter 8: Energization will provide a more detailed discussion of the SFT in the context of initial energization.

4.2.8 *Switching Operations: Precision in Switching Actions*

4.2.8.1 Low-Voltage Switching Operations

Low-voltage switching operations (up to 1000 V) often involve routine tasks like turning supply systems on and off or isolating control lines. Although the voltages are lower, these operations carry risks, especially in the presence of moisture or insufficient grounding. A typical example is disconnecting a transformer's control lines to ensure no control commands are sent to the device.

4.2.8.2 High-Voltage Switching Operations

High-voltage switching operations (over 1000 V) require significantly more elaborate safety measures. Before the switching action, a detailed switching plan is created, describing the sequence of all operations. In the transformer cell, this includes opening circuit breakers, disconnecting grounding lines, and closing disconnectors. Communication occurs via direct connections between offshore and onshore teams to ensure all steps are executed precisely and synchronously.

After each switching operation, tests are conducted to verify the integrity of circuits and the correct operating condition of components. These tests are crucial to confirm the safety and reliability of the entire installation.

Due to the complexity of switching operations—especially in high-voltage areas—a separate chapter is dedicated to this topic. Chapter 8: Energization will detail the requirements, standards, and procedures, including the application of IEC 60364.

4.2.9 *Establishing an Integrated Safety Framework*

The creation of an integrated safety framework is based on documents containing clear guidelines for developing and implementing such systems. This framework combines technical safety measures, organizational processes, and regulatory requirements. The HSE Manager bears the main responsibility for the safety framework, while the Commissioning Manager integrates specific technical aspects, particularly in electrical safety.

A central component of the framework is the regular conduction of risk analyses. The Commissioning Manager contributes technical data and operational insights. For example, regular hazard assessments identify the risks of high-voltage work and define specific control measures.

Additionally, the safety framework is supported by a clear structure of responsibilities (Fig. 4.1). The provided documents emphasize that each team leader is responsible for the safety of their respective department, while the HSE Manager

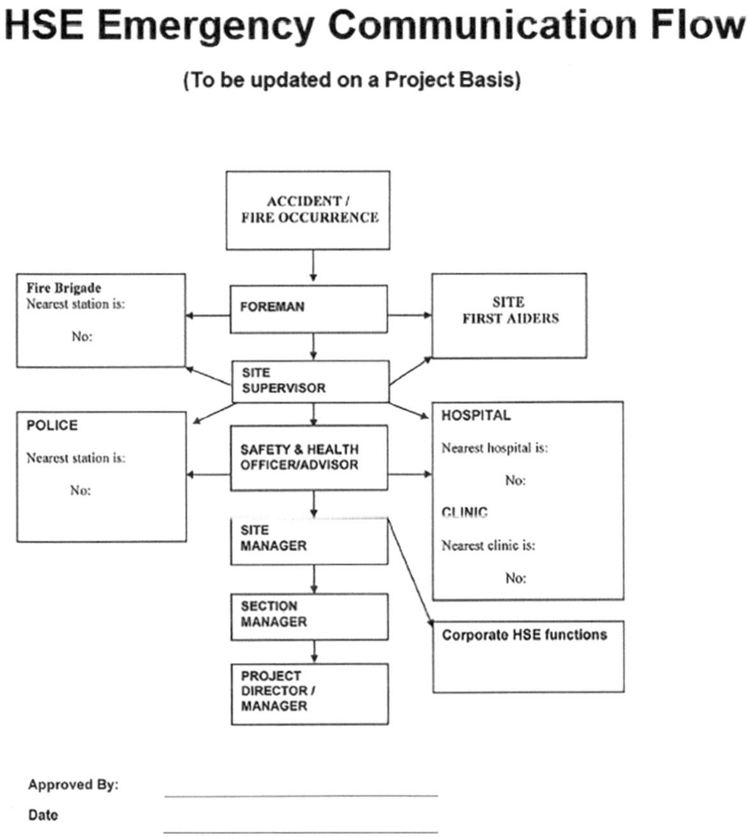

Fig. 4.1 ER flow chart, https://safeworkmethodofstatement.com/emergency-preparedness-response-procedure/, Nov. 2024

acts as an overarching authority. This model promotes clear communication and ensures all project levels are aligned with compliance to safety standards.

4.3 Emergency Management and Lessons Learned

The emergency plans, as described in the provided documents, are comprehensive and cover various scenarios, including medical emergencies, weather-related evacuations, and accidents during electrical work. The HSE Manager creates and maintains these plans, while the Commissioning Manager ensures they address specific electrical risks.

Regular drills, highlighted as an essential part of emergency management, ensure all team members are familiar with the corresponding procedures. For instance, regular evacuation exercises simulate the safe withdrawal from offshore platforms during sudden weather changes.

Lessons Learned are integral to continuous improvement. According to the project's safety guidelines, all incidents are systematically investigated to identify their causes. The HSE Manager conducts these analyses, while the Commissioning Manager contributes technical insights and improvement measures. These findings directly inform updates to safety procedures and optimization of the integrated safety framework.

Safety and risk management in offshore wind projects require precise planning, close collaboration, and the use of specific procedures like LOTO, VoC, and SFT. The cooperation between the HSE Manager and the Commissioning Manager is essential to ensure high standards and secure long-term success.

Chapter 5
Tools and Technology

5.1 Use of Digital Tools for Progress Reporting

In Commissioning Management, progress reports play a crucial role in ensuring transparency and control over complex offshore wind projects. The use of digital tools is revolutionizing this process by enabling precise, real-time monitoring and significantly easing administrative tasks. Progress reports that are created and managed digitally not only enhance efficiency but also minimize sources of error and provide a solid foundation for strategic decisions. This discussion highlights the state of the art in using digital tools in Commissioning Management and shows how they can optimize documentation and communication in offshore wind projects.

5.1.1 The Role of Digital Tools in Progress Management

Traditional progress reports that rely on manual data collection and documentation quickly reach their limits (Fig. 5.1) in complex projects like the commissioning of an offshore wind farm. Digital tools offer decisive advantages by automating data collection, centralizing reports, and ensuring consistent data quality.

Modern software solutions like Project Management Information Systems (PMIS) or specialized plattforms for Commissioning Management—such as Procore or Aconex—integrate data from various sources and present them in clear dashboards. These systems enable seamless communication between offshore and onshore teams and offer real-time insights into project progress. Progress reports created with these tools contain detailed information on tasks performed, open issues, critical milestones, and any delays.

C. Hoppe, *Commissioning Management in Offshore Wind Projects*,
SpringerBriefs in Applied Sciences and Technology,
https://doi.org/10.1007/978-3-032-05358-9_5

INS / CMG / De-CMG / PreC / ReF	Equipment Title	Loc001	Loc002	Loc003	Loc004	Loc005	Loc006	Loc007	Loc008	Loc009	Loc010	Loc011	Loc012	Loc013	Loc014	Loc015	Loc016	Loc017	Loc018	Loc019	Loc020	Loc021	Loc022	Loc023	Loc024	Loc089	Loc090	Loc091	Loc092	Loc093	Loc094	Loc095	Loc096	Loc097	Loc098	Loc099	Loc100	Loc101	Loc102	Loc103	Loc104	Loc105	open (N)	in progress (A)	completed (F)	Comments
Schedule	Cable INS (out)	19.02.20	20.02.20	21.02.20	22.02.20	23.02.20	25.02.20	26.02.20	27.02.20	28.02.20	02.03.20	03.03.20	04.03.20	05.03.20	07.03.20	08.03.20	09.03.20	10.03.20	11.03.20	13.03.20	14.03.20	15.03.20	16.03.20	18.03.20	19.03.20	07.06.20	08.06.20	09.06.20	11.06.20	12.06.20	13.06.20	14.06.20	16.06.20	17.06.20	18.06.20	19.06.20	21.06.20	22.06.20	23.06.20	24.06.20	25.06.20	27.06.20				
Schedule	VTG INS	26.03.2027	28.03.2027	30.03.2027	01.04.2027	03.04.2027	05.04.2027	07.04.2027	09.04.2027	11.04.2027	13.04.2027	15.04.2027	17.04.2027	19.04.2027	21.04.2027	23.04.2027	26.04.2027	28.04.2027	30.04.2027	02.05.2027	04.05.2027	06.05.2027	08.05.2027	10.05.2027	12.05.2027	23.09.2027	26.09.2027	28.09.2027	30.09.2027	02.10.2027	04.10.2027	06.10.2027	08.10.2027	10.10.2027	12.10.2027	14.10.2027	16.10.2027	18.10.2027	20.10.2027	24.10.2027	25.10.2027	26.10.2027				
Schedule	Structure Handover	20.07.2027	20.07.2027	21.07.2027	21.07.2027	22.07.2027	22.07.2027	23.07.2027	23.07.2027	24.07.2027	24.07.2027	25.07.2027	25.07.2027	26.07.2027	26.07.2027	27.07.2027	27.07.2027	28.07.2027	28.07.2027	29.07.2027	29.07.2027	30.07.2027	30.07.2027	31.07.2027	31.07.2027	07.10.2027	10.10.2027	12.10.2027	14.10.2027	16.10.2027	18.10.2027	20.10.2027	22.10.2027	24.10.2027	26.10.2027	28.10.2027	30.10.2027	01.11.2027	03.11.2027	07.11.2027	08.11.2027	09.11.2027				
Schedule	VTG days without grid	102	100	99	97	96	94	93	91	90	88	87	85	84	82	81	78	77	75	74	72	71	69	68	66	0	0	0	0	0	0	0	0	0	0	0	0	0	0	0	0	0				
Schedule	Structure days	276	275	274	273	272	270	270	268	268	266	265	264	263	262	261	259	259	257	256	255	254	253	252	250	214	215	215	216	216	217	217	217	218	218	219	219	219	220	221	222	223				
Installation	Monopile	N	N	N	N	N	N	N	N	N	N	N	N	N	N	N	N	N	N	N	N	N	N	N	N	N	N	N	N	N	N	N	N	N	N	N	N	N	N	N	N	N	105	0	0	
Installation	TP	N	N	N	N	N	N	N	N	N	N	N	N	N	N	N	N	N	N	N	N	N	N	N	N	N	N	N	N	N	N	N	N	N	N	N	N	N	N	N	N	N	105	0	0	
PreChecks	Pre-VTG check week	CW12	CW12	CW13	CW13	CW13	CW14	CW14	CW14	CW14	CW15	CW15	CW15	CW16	CW16	CW16	CW17	CW17	CW17	CW17	CW18	CW18	CW18	CW19	CW19	CW38	CW38	CW39	CW39	CW39	CW40	CW40	CW40	CW40	CW41	CW41	CW41	CW42	CW42	CW42	CW43	CW43				
PreChecks	Pre-VTG	N	N	N	N	N	N	N	N	N	N	N	N	N	N	N	N	N	N	N	N	N	N	N	N	N	N	N	N	N	N	N	N	N	N	N	N	N	N	N	N	N	105	0	0	
Installation	Turbine	N	N	N	N	N	N	N	N	N	N	N	N	N	N	N	N	N	N	N	N	N	N	N	N	N	N	N	N	N	N	N	N	N	N	N	N	N	N	N	N	N	105	0	0	
Commissioning	Diesel Genset																																													
PreChecks	Pre-IAC	N	N	N	N	N	N	N	N	N	N	N	N	N	N	N	N	N	N	N	N	N	N	N	N	N	N	N	N	N	N	N	N	N	N	N	N	N	N	N	N	N	105	0	0	
Installation	Cable connected In	N	N	N	N	N	N	N	N	N	N	N	N	N	N	N	N	N	N	N	N	N	N	N	N	N	N	N	N	N	N	N	N	N	N	N	N	N	N	N	N	N	105	0	0	
Installation	Cable connected Out	N	N	N	N	N	N	N	N	N	N	N	N	N	N	N	N	N	N	N	N	N	N	N	N	N	N	N	N	N	N	N	N	N	N	N	N	N	N	N	N	N	105	0	0	
Installation	FO connection In	N	N	N	N	N	N	N	N	N	N	N	N	N	N	N	N	N	N	N	N	N	N	N	N	N	N	N	N	N	N	N	N	N	N	N	N	N	N	N	N	N	105	0	0	
Installation	FO connection Out	N	N	N	N	N	N	N	N	N	N	N	N	N	N	N	N	N	N	N	N	N	N	N	N	N	N	N	N	N	N	N	N	N	N	N	N	N	N	N	N	N	105	0	0	
Commissioning - Gridless	Turbine	N	N	N	N	N	N	N	N	N	N	N	N	N	N	N	N	N	N	N	N	N	N	N	N	N	N	N	N	N	N	N	N	N	N	N	N	N	N	N	N	N	105	0	0	
Commissioning - Grid	First feed-in	N	N	N	N	N	N	N	N	N	N	N	N	N	N	N	N	N	N	N	N	N	N	N	N	N	N	N	N	N	N	N	N	N	N	N	N	N	N	N	N	N	105	0	0	
Commissioning - Grid	Turbine 240h test	N	N	N	N	N	N	N	N	N	N	N	N	N	N	N	N	N	N	N	N	N	N	N	N	N	N	N	N	N	N	N	N	N	N	N	N	N	N	N	N	N	105	0	0	
Commissioning - Grid	Turbine *TOC*	N	N	N	N	N	N	N	N	N	N	N	N	N	N	N	N	N	N	N	N	N	N	N	N	N	N	N	N	N	N	N	N	N	N	N	N	N	N	N	N	N	105	0	0	
Commissioning	Turbine 500h service	N	N	N	N	N	N	N	N	N	N	N	N	N	N	N	N	N	N	N	N	N	N	N	N	N	N	N	N	N	N	N	N	N	N	N	N	N	N	N	N	N	105	0	0	
Commissioning	Ready for energization - Date of reception																																													
Commissioning - Grid	IAC In energized	N	N	N	N	N	N	N	N	N	N	N	N	N	N	N	N	N	N	N	N	N	N	N	N	N	N	N	N	N	N	N	N	N	N	N	N	N	N	N	N	N	105	0	0	
Commissioning - Grid	IAC Out energized	N	N	N	N	N	N	N	N	N	N	N	N	N	N	N	N	N	N	N	N	N	N	N	N	N	N	N	N	N	N	N	N	N	N	N	N	N	N	N	N	N	105	0	0	
Commissioning - Grid	FO connection In	N	N	N	N	N	N	N	N	N	N	N	N	N	N	N	N	N	N	N	N	N	N	N	N	N	N	N	N	N	N	N	N	N	N	N	N	N	N	N	N	N	105	0	0	
Commissioning - Grid	FO connection Out	N	N	N	N	N	N	N	N	N	N	N	N	N	N	N	N	N	N	N	N	N	N	N	N	N	N	N	N	N	N	N	N	N	N	N	N	N	N	N	N	N	105	0	0	
Commissioning	**TP completed (auto)**	N	N	N	N	N	N	N	N	N	N	N	N	N	N	N	N	N	N	N	N	N	N	N	N	N	N	N	N	N	N	N	N	N	N	N	N	N	N	N	N	N	105	0	0	
Installation	SHM Sensors																																										0	0	0	
Installation	SHM Wave Radar																																										0	0	0	

Fig. 5.1 Excel screenshot of the commissioning list, CHO 2021

5.1.2 Automated Data Collection and Reporting

Digital tools reduce the effort required for data collection by integrating information directly from various sources. Sensors on offshore wind turbines and substations, for example, provide real-time data on technical tests, energy status, or system faults. This data is automatically fed into the software, analyzed, and prepared in standardized report formats.

An example of this is the integration of Supervisory Control and Data Acquisition (SCADA) systems (Fig. 5.2) into the reporting tool. These systems monitor electrical and mechanical components and transfer the collected data directly to the reporting system. Automated reports ensure that no relevant information is missing and that data consistency is maintained.

Another advantage is the automation of progress evaluation. Digital tools use algorithms to calculate project progress in percentages based on completed tasks and display this information in a Gantt chart. This facilitates visual tracking of progress and allows for quick identification of delays.

5.1.3 Real-Time Dashboards and Visualizations

Progress reports are significantly enhanced through the use of dashboards. These dashboards provide a visual representation of the current project status and enable Commissioning Management to immediately recognize critical areas. For example, a dashboard powered by real-time data from SCADA systems shows the status of switching operations, the progress of commissioning individual turbines, or the availability of resources.

Such visualizations can be presented in the form of charts, maps, or tables, which can be flexibly adjusted to meet specific requirements. An example is displaying the progress of Factory Acceptance Tests (FAT) and Site Acceptance Tests (SAT). A color-coded scheme shows which tests have been completed, which are still pending, and where potential risks exist.

An additional advantage of real-time dashboards is the ability to create simulations and scenarios. Commissioning Management can simulate alternative schedules or resource allocations to assess the impact of changes in the project plan.

5.1.4 Communication and Collaboration

Digital tools promote collaboration among various stakeholders by providing a central platform for information exchange. Progress reports can be shared in real-time via cloud-based systems, allowing all participants—from offshore technicians to onshore managers—to have access to up-to-date information at any time.

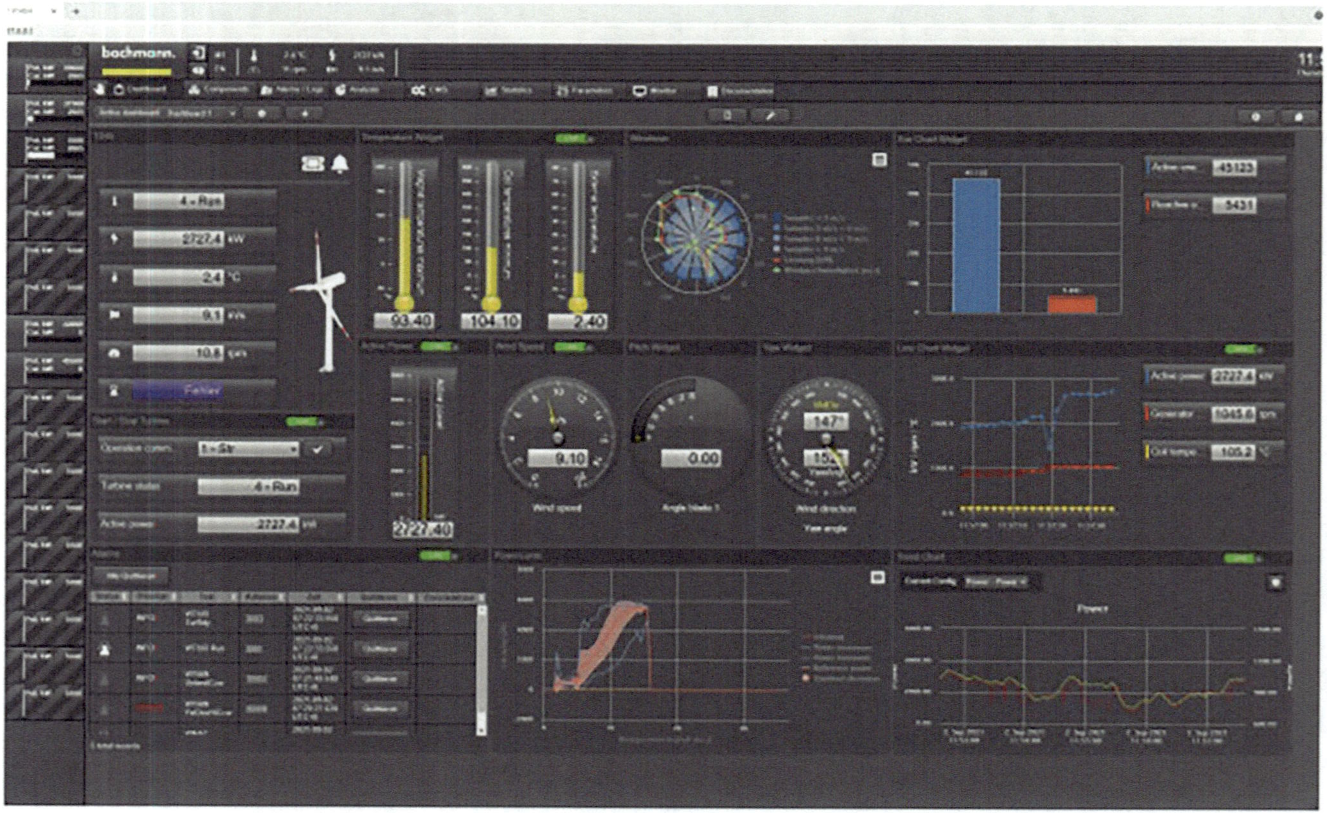

Fig. 5.2 Bachmann SCADA GUI, https://www.bachmann.info/de/branchen/energie/wind-energie/webtechnologie, Nov. 2024—Copyright by Bachmann electronic GmbH, www.bachmann.info

Particularly useful is the ability to comment on reports and track changes directly in the system. For example, technicians in the offshore sector can add specific comments on tested components, which are then automatically integrated into the central report. This transparency facilitates coordination between teams and promotes efficient decision-making.

Furthermore, digital tools enable the integration of notification systems. Commissioning Management is automatically informed when a milestone is reached or when critical tasks are delayed. This proactive communication improves responsiveness and minimizes potential project risks.

5.1.5 Examples of Advanced Digital Tools

Some of the most advanced tools used in Commissioning Management are:

- **Primavera P6**: This platform enables the creation of detailed schedules and their integration into progress reports. The tool offers a seamless connection between Gantt charts and real-time data sources.
- **Procore**: A cloud-based solution that combines progress reporting with document management and resource planning. Procore also supports integration with third-party software like SCADA systems.
- **BIM 360**: Building Information Modeling (BIM) tools enable the integration of 3D models into the progress report. This is particularly helpful for visualizing the status of individual components within an offshore substation.
- **LAUTEC**: Lautec (Fig. 5.3) is a specialized software platform developed for planning, monitoring, and optimizing processes in offshore wind projects. Focusing on real-time data management and transparency, Lautec supports Commissioning Management in centrally managing progress reports, resource plans, and documentation. Especially for the harmonization of Daily Progress Reports (DPR), Lautec is a user-friendly software solution.

The topic of DPR and reporting in general is discussed in more detail in Part 4, Chap. 11.

5.1.6 Challenges and Future Developments

Despite the numerous advantages, there are challenges in implementing digital tools in Commissioning Management. A common hurdle is the integration of different data sources, as many offshore systems use proprietary interfaces. Additionally, effective use of such tools requires comprehensive training of staff to ensure they are familiar with the systems and can use them efficiently.

In the future, it is expected that artificial intelligence (AI) and machine learning will play an even greater role in progress management. AI-supported systems could

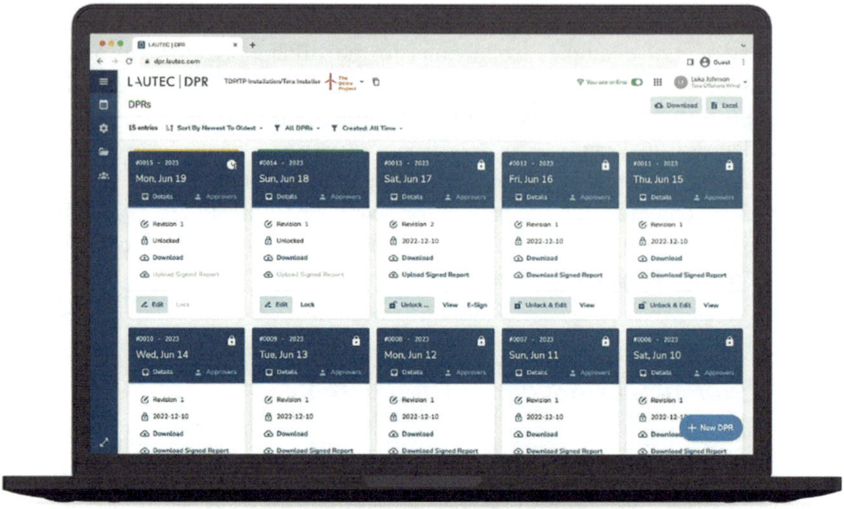

Fig. 5.3 Lautec DPR, https://lautec.com/2024/04/17/dpr-and-cadeler/, Nov. 2024—Copyright by Lautec

make predictions about delays by analyzing historical data and provide automatic recommendations for optimizing schedules. Such developments have the potential to further revolutionize Commissioning Management and make projects even more efficient.

5.1.7 Conclusion

The use of digital tools for progress reporting has fundamentally changed Commissioning Management. It not only enables more precise and efficient documentation but also promotes collaboration and improves transparency throughout the project. By integrating advanced technologies, the management of complex offshore wind projects is optimized, and the risk of delays and errors is minimized. In an industry increasingly characterized by technological innovations, digital progress reports are an indispensable tool for any successful project.

5.2 Software for Audits and Handover Processes

Audits and handover processes are critical phases in Commissioning Management that ensure transparency, traceability, and compliance with quality standards. The use of specialized software solutions has revolutionized these processes in recent years by automating manual tasks, optimizing data flows, and creating a central platform for all participants. This section illustrates how software tools support audits and handover processes in Commissioning Management and what advantages they offer, particularly in the complex environment of offshore wind projects.

5.2.1 *The Role of Software in Audits*

Audits are an integral part of Commissioning Management as they ensure compliance with technical, safety, and regulatory requirements. Software tools for audits provide a structured and standardized platform to conduct inspections efficiently and transparently. A typical scenario is the audit of an offshore substation, where components such as transformers, cable systems, and safety equipment are inspected.

5.2.2 *Digital Checklists and Automation*

Modern audit software like iAuditor or PlanRadar allows for the creation of digital checklists tailored to the specific requirements of a project. These checklists contain detailed inspection points for each component or process, including visual inspections, functional tests, and safety checks. By capturing data digitally, inspectors can enter information directly on-site, often via mobile devices like tablets or smartphones.

Automation significantly reduces administrative effort. Audit reports are automatically generated once the checklists are completed and can be shared in real-time with other project participants. This automation ensures that no inspection points are overlooked and that all relevant information is centrally stored.

5.2.3 *Issue Tracking and Follow-Up*

Software tools often integrate modules for issue tracking and follow-up, allowing non-conformities to be documented immediately and corrective actions to be initiated. An example is the audit of a transformer bay where a faulty relay is discovered. The software records the issue, assigns it to the responsible person, and automatically sets a deadline for resolution. Progress in rectifying the issue can also be tracked in the software, increasing transparency.

5.2.4 Integration of Reporting Standards

Many audit software solutions support industry-specific standards like ISO 9001 or IEC 60364, which is particularly important for offshore wind projects subject to strict regulatory requirements. The software ensures that all audits comply with the corresponding standards and provides templates that simplify documentation.

5.2.5 Handover Processes: Efficiency Through Digital Tools

The handover process marks the transition from the commissioning phase to the operational phase and requires precise documentation of all work performed and tests conducted. The use of software in this area ensures that this transition occurs smoothly, efficiently, and is thoroughly documented.

5.2.6 Central Platform for Documentation

Software solutions like Aconex or Procore provide a central platform where all relevant documents for the handover process are stored and managed. These platforms allow for the consolidation of test reports, certificates, inspection data, and other important documents into a single system. Central storage ensures that all participants—from technicians to operations managers—can access the necessary information at any time.

An example is the handover of an offshore turbine: The software documents all conducted tests, such as mechanical inspections and electrical commissioning tests, and creates a complete handover report that must be approved by all parties before the turbine is officially put into operation.

5.2.7 Automatic Handover Packages

An outstanding feature of modern software is the ability to automatically create handover packages. These packages include all necessary documents and information required for acceptance. Instead of assembling the documents manually, the software automatically generates a complete package containing test reports, technical drawings, certificates, and other relevant documents.

5.2.8 Digital Approval Workflows

Handover software integrates digital approval workflows that allow handover documents to be reviewed and released electronically. This reduces time consumption and improves traceability, as every approval is logged in the system. For example, operators can add comments or questions on specific points during the handover process, which are then reviewed and addressed by Commissioning Management.

5.2.9 Integration of Audits and Handover Processes

Another advantage of modern software tools is the integration of audits and handover processes into a single platform. This integration enables a seamless handover, as all data captured during the audit process is automatically transferred into the handover process.

A practical example is the commissioning of an offshore substation. During the audit, all relevant inspection points are documented, and non-conformities are directly transferred into the handover system. Once these points are resolved, the handover package is automatically updated, significantly accelerating the process.

5.2.10 Real-Time Communication and Collaboration

Cloud-based software solutions enable real-time communication between offshore and onshore teams as well as other stakeholders. All participants can view progress, documentation, and changes directly in the software, making the handover process more transparent and efficient. Notification functions inform Commissioning Management about changes or pending approvals, minimizing delays.

5.2.11 Conclusion

The use of software for audits and handover processes has significantly improved the efficiency and transparency of these critical phases in Commissioning Management. By automating tasks, standardizing documentation, and facilitating collaboration, these tools help ensure that offshore wind projects meet the highest quality and safety standards. In an industry where precision and reliability are paramount, leveraging advanced software solutions is essential for successful project completion.

Part III
Practice in Commissioning

The practical application of methods, processes, and standards is the key to the successful execution of a commissioning project. In Part III, we show how the previously described fundamentals, processes, and tools are applied in real offshore wind projects. The focus is on practical approaches to overcoming typical challenges, such as coordinating complex switching operations, the safe energization of low-voltage components, and compliance with technical and regulatory standards. Through detailed descriptions of processes like batch energization and initial energization, proven solutions are presented that ensure efficiency and safety. This section combines solid theory with concrete application scenarios and offers valuable insights for the successful commissioning and operation of offshore wind farms.

Chapter 6
Commissioning of Components

6.1 Wind Turbine Generators (WTGs)

6.1.1 Responsibilities and Task Distribution

The commissioning of Wind Turbine Generators (WTGs) is usually carried out directly by the respective wind turbine manufacturer. This practice ensures that all tests and inspections are performed by experts who are thoroughly familiar with the specific technical requirements and characteristics of the particular turbine types. Therefore, the main task of the commissioning management of the offshore wind project lies less in the direct execution of the commissioning but more in supervising and coordinating interfaces, as well as in the preparation and follow-up of the required documents.

6.1.2 Interface Management and Documentation

Interface management is a central component of the commissioning management tasks. It ensures that all prerequisites for the commissioning of the WTGs are met. This includes providing a safe and accessible installation site, the availability of necessary energy and connections, and the coordination of schedules between different trades.

Another important aspect is the review and post-processing of documentation created by the wind turbine manufacturer. This includes protocols of conducted tests, acceptance certificates, and technical reports. The commissioning management checks these documents for completeness and compliance with project-specific requirements to ensure a smooth handover to operations.

C. Hoppe, *Commissioning Management in Offshore Wind Projects*,
SpringerBriefs in Applied Sciences and Technology,
https://doi.org/10.1007/978-3-032-05358-9_6

6.1.3 Coordination of the Site Acceptance Test (SAT)

The Site Acceptance Test (SAT) represents a central milestone in the commissioning process of the WTGs. This test is planned and conducted in close collaboration with the wind turbine manufacturer. The commissioning management ensures that all requirements and inspections of the SAT correspond to the project-specific specifications and are integrated with the overall system of the offshore wind farm.

An important part of the SAT is the verification of the central control unit of the wind turbines, often referred to as the Park Master or Park Pilot. This control unit is responsible for the overarching control and regulation of the entire turbine fleet. As part of the SAT, functions such as load distribution, control and regulation capability, and communication with the offshore substation and control systems are tested.

6.1.4 Control and Regulation Capability

Another central test point is the control and regulation capability of the WTGs. Here, it is ensured that the turbines can respond to all necessary commands, including rotor speed regulation, load limitation, and adaptation to grid requirements. These tests are essential to ensure the optimal performance of the WTGs and their integration into the overall system of the wind farm.

6.1.5 Collaboration and Quality Assurance

Close collaboration between the wind turbine manufacturer and the commissioning management is crucial to make the commissioning process efficient and safe. Regular coordination, joint progress reports, and the review of test results ensure that all components work together smoothly and meet the high standards for offshore wind projects.

Thus, the commissioning management assumes a supportive and coordinating role that focuses on interface supervision, document review, and ensuring system integration. This approach ensures the successful completion of the commissioning phase and the handover of the WTGs to commercial operation.

6.2 Inner Array Grid (IAG) and Cables

The Inner Array Grid (IAG) forms the backbone of an offshore wind farm by connecting the individual wind turbine generators (WTGs) to the offshore substation. The commissioning of this network requires careful planning and execution of specific tests to ensure the integrity and performance of the cable systems.

6.2.1 Cable Installation and Inspection

The installation of the IAG cables begins with the careful laying and protection of the cables to avoid mechanical damage. After laying, a visual inspection is carried out to ensure that there are no obvious damages or irregularities.

6.2.2 Electrical Tests

After the mechanical inspection, electrical tests are essential to ensure the functionality and safety of the cables. Common test procedures include:

- **Insulation Resistance Test**: This test measures the resistance of the cable insulation to ensure that no unwanted leakage currents occur. A high insulation resistance indicates intact insulation.
- **High Voltage Test (VLF Test)**: The Very Low Frequency (VLF) test is applied to check the voltage withstand capability of the cables. A very low frequency alternating voltage is applied to simulate the stress on the insulation under operating conditions. It should be noted that this form of cable testing is rarely used in projects nowadays. This is mainly due to the now high voltages of 66 kV and above. The VLF test equipment becomes simply too unwieldy when testing with $3 \times U_0$, i.e., $66\ \text{kV}/\sqrt{2} \times 3 = 140\ \text{kV}$.
- **Damped Alternating Current (DAC) Test**: DAC is a current alternative to the VLF test. This method uses damped alternating voltages for the diagnosis of power cables. It enables the detection of partial discharges and the assessment of the condition of the cable insulation. The DAC test is particularly effective for quality control of newly installed high-voltage cable connections and supports condition-based maintenance throughout the cable's lifetime.
- **Partial Discharge Measurement**: This method detects defects or weak points in the cable insulation that can lead to partial discharges. Early detection allows for preventive measures to avoid failures.
- **Time Domain Reflectometry (TDR)**: This procedure can locate positions of irregularities or faults along the cable. An impulse is sent into the cable, and reflected signals are analyzed to identify disturbances.

6.2.3 *Connection Tests and Verification of Junction Boxes*

The junction boxes that connect the cables between the WTGs and the substation must also be thoroughly tested. This includes:

- **Continuity Test**: Ensuring that electrical connections have been correctly made and that there are no interruptions.
- **Grounding Test**: Verifying proper grounding to ensure the safety and functionality of the system.

6.2.4 *Load Tests and Performance Monitoring*

After the completion of mechanical and electrical tests, the entire IAG is tested under load conditions. This serves to verify the system's ability to transport the generated energy efficiently and without losses to the offshore substation. Monitoring systems, such as SCADA, provide real-time data on system performance and help detect potential problems early.

By carefully conducting these tests and inspections, it is ensured that the Inner Array Grid meets the high requirements for reliability and efficiency in offshore wind farms.

6.3 Offshore Substations and Control Systems

The offshore substation is the central link between the wind turbine generators and the transmission grid. Its commissioning requires a highly structured approach, as a multitude of systems—from primary and secondary components to control systems—must be coordinated. The commissioning manager is responsible for setting priorities, managing dependencies, and coordinating the individual test phases: Factory Acceptance Test (FAT), Harbour Acceptance Test (HAT), Offshore Commissioning, and Site Acceptance Test (SAT). These prioritizations ensure that all systems are tested and commissioned in the optimal sequence.

6.3.1 *Factory Acceptance Tests (FAT): Quality Assurance Before Installation*

The FAT phase serves to verify the functionality of the primary systems before they are transported to the offshore substation. Primary systems include transformers, switchgear, and protection relays, which are crucial for the transmission and safety of high-voltage power.

During the FAT, these systems are tested under controlled conditions. Examples include:

- **Transformers**: Insulation and voltage withstand tests, as well as thermal performance tests to ensure efficiency and safety.
- **Switchgear**: High-voltage tests that check the switching capacity and ability to disconnect loads.
- **Protection Relays**: Functional tests using fault current simulations to ensure the relays trip correctly.

The FAT provides the opportunity to detect and rectify problems early before the components are installed offshore. The commissioning manager coordinates the FAT results with the project team to ensure that all specifications are met and the components are ready for transport.

6.3.2 Harbour Acceptance Tests (HAT): Interface Between Production and Offshore Operation

The Harbour Acceptance Test (HAT) is a crucial step in transitioning from the production phase to offshore installation. During the HAT, it is verified that the systems remain functional after their transport to the port facilities and that all interfaces work correctly. The focus here is on integrating the primary systems with the supporting infrastructures.

Typical tests include:

- **System Connections**: Ensuring that electrical connections are correctly installed and intact.
- **Mechanical Tests**: Checking fastenings and structures, especially after transport vibrations.
- **Communication Tests**: Initial verification of communication interfaces between primary systems and control systems.

The commissioning manager prioritizes systems critical for offshore installations, such as emergency power supply or lighting systems. This guarantees that the substation is ready for operation immediately after installation

6.3.3 Offshore Commissioning: Integration and Prioritization of Systems

Offshore commissioning is the most complex phase, as all systems of the offshore substation are tested and integrated in a real operating environment. The prioritization of systems is of utmost importance, as some systems, like emergency lighting and the UPS system, are necessary for the safety and functionality of other components.

The matrix shows that some systems necessarily require certain predecessors. For example:

- **Low Voltage Systems**: These must be installed and tested before all electrical systems, as they serve as the basis for most control and protection systems.
- **High Voltage Systems**: These require prior installation and testing of grounding systems and lightning protection systems to minimize safety–critical risks.
- **SCADA and Control Systems**: These can only be integrated once all primary systems are functional, as they act as the overarching control level.

The commissioning manager sets the priorities so that critical systems like lighting, safety, and communication are tested first. This enables a safe environment for subsequent tests and reduces the risk of delays.

In addition to the main systems of an offshore substation, subsystems play a crucial role in ensuring the safe operation and functionality of the entire installation. Prioritizing these subsystems during commissioning is essential, as they often form the foundation for other tests and system integrations.

- **HVAC Systems**: Climate control ensures that sensitive electrical and electronic equipment can operate under optimal environmental conditions and is therefore a prerequisite for subsequent commissioning tests.
- **Firefighting Systems (Foam and Gas)**: These systems ensure fire protection in critical areas and must be operational before commissioning transformers, reactors, and other safety-relevant components.
- **Communication and Alarm Systems (PAGA)**: These systems are crucial for efficiently disseminating information in emergencies and ensuring the safety of offshore personnel.
- **Wastewater Systems and Oil Separators**: These systems minimize environmental risks by effectively handling oily wastewater and preventing harmful substances from entering the sea.

Prioritizing and integrating these subsystems is vital to ensure safety and the smooth execution of offshore commissioning. The commissioning manager determines the sequence in which these subsystems are tested to optimally prepare subsequent steps.

6.3.4 Site Acceptance Tests (SAT): Release for Operation

The Site Acceptance Test (SAT) represents the final step, in which all systems of the offshore substation are tested for functionality and integration. These tests are conducted under real operating conditions and include:

- **End-to-End Communication**: Verification of the complete communication chain from the wind turbine generators to the substation and further to the onshore control center.
- **Grid Simulation**: Tests simulating various grid loads to assess the stability and efficiency of the substation.
- **Redundancy Checks**: Ensuring that backup systems like emergency power supply function correctly.

The commissioning manager closely coordinates the SAT with the involved teams to ensure that all test results are documented and the substation is approved for commercial operation.

6.3.5 Coordination and Prioritization: The Role of the Commissioning Manager

The matrix illustrates that coordination and prioritization of systems is a central responsibility of the commissioning manager. By clearly defining sequences and dependencies, they ensure that commissioning proceeds efficiently and safely. The manager must also respond to unforeseen challenges, such as delays in a critical system, and establish alternative prioritizations.

An example: If the installation of the high-voltage system is delayed, the manager can prioritize alternative tests, such as the integration of control systems, to avoid extending the overall commissioning duration.

Commissioning offshore substations and control systems requires a high degree of planning, coordination, and prioritization. By clearly defining sequences and carefully coordinating between FAT, HAT, offshore commissioning, and SAT, the commissioning manager ensures that all systems work together smoothly and meet the high demands of an offshore wind project. The use of structured matrix approaches and consideration of critical dependencies are crucial for success.

Chapter 7
Troubleshooting and Fault Resolution

Troubleshooting within the scope of commissioning management for offshore wind projects is an essential task to ensure the reliability and operational readiness of complex systems. The specific challenges presented by offshore environments—such as limited accessibility, high safety requirements, and variable weather conditions—demand a systematic, well-documented approach to identifying and resolving issues. This chapter outlines a step-by-step methodology for fault analysis and presents case studies that highlight typical problems and proven solutions.

7.1 Step-By-Step Guides for Fault Analysis

Effective troubleshooting in offshore wind projects is based on a seven-step approach that ensures safety, saves time, and complies with regulations. The following steps provide a structured methodology for analyzing and resolving faults:

1. **Gather Information**

   The first step involves collecting all relevant information about the affected system or component. This includes technical documentation, operational histories, and previous issues and their solutions. Important questions include:

   - What are the operating parameters of the system?
   - Have similar problems occurred in the past, and how were they resolved?
   - Are there environmental factors (e.g., high humidity, salt exposure) that could impair function?

C. Hoppe, *Commissioning Management in Offshore Wind Projects*,
SpringerBriefs in Applied Sciences and Technology,
https://doi.org/10.1007/978-3-032-05358-9_7

2. **Analyze the Fault**
   Once sufficient information is available, investigate how the system deviates from its intended function. For example, with a high-voltage transformer, check whether the cooling system is operating properly or if unusual voltage fluctuations occur.
3. **Define Parameters to be Tested**
   Next, determine which parameters need to be measured and evaluated. In electrical systems, these could be voltage, current flow, or insulation resistance. It is important to identify the required measuring equipment and ensure that the measurement points are safely accessible.
4. **Narrow Down the Fault Source**
   The faulty component or affected subsystem is identified through targeted tests and elimination methods. For instance, if a circuit breaker keeps tripping, check whether the cause lies in the connected load, the breaker itself, or the protection settings.
5. **Resolve the Fault**
   After identifying the cause, repair or replace the affected component. In the offshore environment, special measures such as manufacturer support may be required. Strict adherence to safety regulations, such as Lock-Out Tag-Out (LOTO), is essential during the repair.
6. **Verify the Repair**
   After resolving the fault, conduct comprehensive tests to ensure that the problem has been fully addressed and the system meets specifications again. Simulations or trial runs under realistic conditions can help here.
7. **Root Cause Analysis**
   The final step is a detailed root cause analysis to identify the fundamental triggers of the problem. The results are documented to prevent future faults and serve as a basis for continuous project improvement.

Fault analysis in offshore commissioning follows these seven systematic steps. Each step requires careful documentation to ensure transparency and traceability. Strict adherence to safety protocols like LOTO is essential throughout the fault analysis. This systematic approach minimizes downtime and ensures the safe and efficient commissioning of systems.

7.2 Case Studies of Typical Problems and Their Solutions

7.2.1 Case Study 1: Failure of a Transformer Cooling System

During commissioning, the cooling system of a high-voltage transformer failed. Initial tests showed that the fan mechanics were intact. A more in-depth analysis revealed that the control relay of the fans was damaged by salt deposits. The relay was replaced, and the affected components were additionally protected against corrosion. The lesson learned was the need to improve the protection of exposed electrical equipment from salty air.

7.2.2 Case Study 2: Communication Loss in the SCADA System

An offshore substation reported sporadic communication outages in the SCADA system. Analysis of the signals revealed that electromagnetic interference from inadequately shielded cables near high-voltage equipment was the cause. The cables were rerouted and provided with additional shielding, eliminating the disturbances.

7.2.3 Case Study 3: Tripping of a Circuit Breaker

A circuit breaker in the low-voltage distribution panel repeatedly tripped during load testing. Investigation showed that the thermal protection setting of the breaker was not adjusted to the expected load. After recalibration, the system functioned flawlessly. This underscored the importance of careful configuration checks during commissioning.

These case studies demonstrate that effective troubleshooting is based on systematic analysis and targeted actions to efficiently resolve issues. Clear diagnostic steps are crucial, such as identifying environmental factors (e.g., corrosion), using appropriate testing methods (e.g., signal verification), and adjusting system settings. In each case, close collaboration between technicians, commissioning teams, and specialists was emphasized to tackle specific challenges like communication failures or protection mechanisms. A thorough root cause analysis after fault resolution was essential to develop preventive measures and sustainable solutions. The importance of precise documentation and strict adherence to safety standards was consistently highlighted. Together, the case studies underscore that systematic planning, technical expertise, and cross-team communication are the foundation for successful troubleshooting in the offshore environment.

7.3 Particularities and Challenges in Offshore Troubleshooting

Offshore troubleshooting requires a special approach because the complex conditions and potential hazards of an offshore environment go far beyond standard requirements. Safety is always the top priority, as limited accessibility, adverse weather conditions, and proximity to high-voltage and high-power systems pose significant risks. To meet these challenges, strict safety protocols and dynamic adaptation to on-site conditions are essential.

A crucial aspect is the necessity to create or adapt individual risk assessments and method statements directly on-site. Standard templates are often insufficient, as troubleshooting brings unpredictable problems not accounted for in usual workflows. Templates can serve as a basis but must be specifically tailored to the situation. The assessment of risks and the planning of measures are carried out in close coordination with the HSE Manager (Health, Safety and Environment), who ensures that all work complies with applicable regulations and safety standards.

A particular challenge is the ad-hoc creation of work permits (Permit-to-Work, PtW). Since troubleshooting often requires unplanned work, PtWs must be issued at short notice and considering all other ongoing activities. This ensures that no risks arise from overlaps with other teams, systems, or work. For example, it must be checked whether adjacent systems are de-energized or whether neighboring work could be affected by the troubleshooting. These dynamic permits require close collaboration between the troubleshooting team and coordinators of other work areas.

Safety protocols like Lock-Out Tag-Out (LOTO) play a central role in offshore troubleshooting. They guarantee that all energies—whether electrical, mechanical, or hydraulic—are safely isolated before work begins. Especially during fault finding, where systems are often operated or tested in non-standard states, it must be ensured that no uncontrolled energy supply occurs.

Another important aspect is documentation and clear communication. In the offshore environment, multiple teams often work in parallel on different tasks, requiring clear coordination. Troubleshooting must be organized so that all affected teams are informed in good time about possible impacts. Ad-hoc meetings and continuous updating of work plans are essential. This not only ensures transparency but also prevents potential conflicts or safety risks due to uncoordinated work processes.

Preparation for unforeseen challenges is also essential. Offshore operations require precise planning and logistics, as spare parts and special tools are often only available with significant effort. Therefore, troubleshooting teams must be equipped with sufficient resources to respond flexibly to various scenarios. Prepared emergency plans and alternative strategies ensure that unexpected problems do not massively delay workflows.

Furthermore, specialized tools and measuring devices are required that can withstand the harsh conditions of the offshore environment. These devices must be regularly maintained and certified for use in explosive or humid environments. Their use requires special training of personnel to ensure they are used effectively and safely.

In summary, troubleshooting in the offshore sector requires a flexible, dynamic approach with safety, precise planning, and real-time adaptation at the forefront. The combination of individually tailored safety measures, close collaboration with HSE management, and transparent communication enables efficient and safe fault resolution and maintenance of operations.

Chapter 8
Energization

The commissioning of electrical systems, also known as energization, is a crucial milestone in offshore wind project management. It involves the systematic planning, coordination, and execution of switching operations to ensure that installations can be safely and efficiently supplied with power. Especially in offshore projects, this requires a high level of planning due to environmental influences, technical complexity, and the involvement of various stakeholders.

8.1 Energization of Low-Voltage Components

The energization of low-voltage components is a critical step in the commissioning of offshore wind farms. This process requires strict adherence to international and national standards to ensure the safety and reliability of electrical installations. Essential standards in this context include the NORSOK Standard E-001, the German standard DIN VDE 0100-600, and the international standard IEC 60364.

8.1.1 Relevant Standards and Their Requirements

1. **NORSOK Standard E-001**: This standard sets requirements for electrical installations in offshore facilities and aims to ensure safety in the design and operation of electrical systems. It includes provisions for the selection and use of electrical equipment for the generation, storage, distribution, and utilization of electrical energy in offshore units.
2. **DIN VDE 0100-600**: This German standard, based on IEC 60,364-6, defines the requirements for the initial testing of electrical installations. It details procedures

C. Hoppe, *Commissioning Management in Offshore Wind Projects*,
SpringerBriefs in Applied Sciences and Technology,
https://doi.org/10.1007/978-3-032-05358-9_8

for inspection, testing, and measurement to ensure the safety and functionality of low-voltage systems.

3. **IEC 60364**: As an international standard, IEC 60364 provides comprehensive guidelines for the design, erection, and verification of electrical low-voltage installations. It serves as the basis for many national standards, including the DIN VDE 0100 series.

8.1.2 Procedure for Energization in Compliance with Standards

1. **Planning and Preparation**
    - **Documentation**: According to DIN VDE 0100-600, all relevant documents, including circuit diagrams and test reports, must be complete and up-to-date before commissioning.
    - **Risk Assessment**: NORSOK Standard E-001 emphasizes the need for a comprehensive risk assessment to identify and minimize potential hazards during energization.
2. **Execution of Energization**
    - **Visual Inspection**: Before energization, DIN VDE 0100-600 requires visual inspections to ensure that the installation meets safety requirements.
    - **Measurements and Tests**: IEC 60364-6 mandates specific measurements, such as verifying insulation resistance and protective conductor connections, to ensure electrical safety.
    - **Gradual Energization**: NORSOK Standard E-001 recommends the gradual energization of low-voltage components to monitor system stability and detect potential issues early.
3. **Safety Measures**
    - **Isolation and Labeling**: According to DIN VDE 0100-600, all parts of the installation to be tested must be isolated and clearly labeled before energization to prevent unintended connections.
    - **Earthing and Short-Circuiting**: IEC 60364 stresses the importance of earthing and short-circuiting conductors during tests to ensure personnel safety.

Energizing low-voltage components in offshore wind farms requires strict compliance with relevant standards. While it's fundamentally open which norms and standards are used and considered in contracts and procedures, careful planning, execution, and monitoring in accordance with these standards ensure the safety and reliability of electrical installations and personnel.

8.2 Planning of High-Voltage Switching Operations

8.2.1 Interface Management with Project Participants

Planning switching operations requires close coordination with all relevant project participants, including the cable installer, the turbine manufacturer, and other technical trades. These interfaces are essential because, during the commissioning phase, work often continues on cable installations (Fig. 8.1), protection systems, or turbine controls. To ensure the safety of all involved, clear points of separation must be defined and appropriately labeled. These separation points ensure that no energy flows occur in areas where work is still in progress.

An example is the completion of Inner Array Grid (IAG) cable installations, which often conclude parallel to the commissioning of the offshore substation. The cable installer must guarantee that all affected cables are properly tested and released before being connected to the electrical system. The same applies to the turbine manufacturer, whose internal switchgear and control systems in the turbines might already be operational while the external grid connection is still being established.

Fig. 8.1 Offshore substation with cables, https://www.nexans.us/en/newsroom/news/details/2022/12/Nexans-AmerCable-awarded-offshore-substation-cable-contract-for-Revolution-Wind Farm-project.html, Nov. 2024—Copyright by Nexans

8.2.2 Sequential Disconnection and Connection

Sequential disconnection and connection require careful technical and logistical planning. Technically, there are often circuit breakers only on the offshore substation intended for switching IAG cables. The switchgear panels in the wind turbines often contain circuit breakers designed for the respective turbine but do not have a direct function for switching entire strings.

These technical conditions necessitate precise coordination to ensure that individual strings can be safely isolated and energized without affecting other systems. Another important point is the sequence of energization, which is determined by several criteria:

1. **Technical Priority**: Strings that have been fully tested and released are energized first.
2. **Logistical Factors**: The availability of technicians, verification of physical access to installations, and weather conditions influence the sequence.
3. **Grid Compatibility**: Coordination with the grid operator is essential to avoid overloading or destabilizing the grid.

8.2.3 Batch Energization

Batch energization refers to the simultaneous connection of multiple systems or components, primarily concerning wind turbines (Fig. 8.2). This method is particularly efficient but presents specific challenges. A central aspect is the lack of reactive power compensation that occurs with IAG cables. This uncompensated reactive power can significantly impact grid stability and requires careful coordination with the grid operator.

Efficiency and grid compatibility often conflict here. While simultaneous connection of components saves time and resources, it increases the risk of voltage instabilities or overloads. To resolve this conflict, a gradual connection combined with temporary compensation measures is often applied. Close monitoring of grid parameters and continuous communication with the grid operator are indispensable.

8.2.4 Weather and Environmental Conditions

Weather influences pose a significant challenge in offshore work. Particularly wind speeds, visibility conditions, and sea state can affect technician safety and the stability of installations. Therefore, switching operations must be planned for time windows with optimal weather conditions. Additionally, the use of weather forecasts requires constant adjustment of the schedule, as offshore conditions can change rapidly.

Fig. 8.2 WTGs offshore, by Orsted, https://seawanderer.org/orsted-chooses-sgres-turbines-for-its-offshore-wind-power-projects, Nov. 2024

8.2.5 Grid Interruptions and Revenue Losses

Energizing installations often involves temporary grid interruptions. These not only lead to revenue losses but can also impact the safety of installations and the functioning of systems already in operation. Close coordination with the grid operator and precise planning of downtime are therefore indispensable to minimize economic and operational impacts.

8.2.6 Qualification of Switching Personnel

Performing switching operations requires highly qualified personnel. Authorized Persons (APs) must be thoroughly trained, especially regarding the specific requirements of offshore installations. Their qualification encompasses not only technical knowledge but also the ability to recognize potential hazards early and respond safely to unforeseen events.

8.3 Switching Operations in Practice

Practical switching operations are crucial for the safe and efficient commissioning of offshore wind farms. They require clear role assignments, precise communication, and strict adherence to standardized processes. Especially the initial energization of new installations brings challenges that are managed through the use of structured procedures and a **Switching Order**. This section details how the various roles—from the Senior Authorized Person (SAP) to the Authorized Person (AP)—and the Operation Control Center (OCC) are involved in the switching process.

8.4 Roles and Responsibilities

8.4.1 Senior Authorized Person (SAP)

The SAP bears direct responsibility for the switchgear and all electrical work within a defined area. They are responsible for creating and approving the Switching Order, including reviewing all safety measures and coordinating with the OCC. The SAP ensures that all involved personnel are informed about the planned switching operations and understand their tasks. During the switching operation, the SAP monitors correct execution and serves as a point of contact for the AP and the OCC.

8.4.2 Authorized Person (AP)

The AP carries out the actual switching operations and is responsible for ensuring that these are performed according to the Switching Order and in compliance with all safety regulations. Before starting the switching operation, the AP verifies their own qualifications, the condition of the switching devices (Fig. 8.3), and the availability of all necessary tools. During execution, they document each step and maintain continuous communication with the OCC.

8.4.3 Operation Control Center (OCC)

The OCC plays a central role in coordinating and monitoring switching operations. It serves as the communication interface between on-site personnel and central control. The OCC conducts the final review of the Switching Order, logs the entire switching operation, and responds to anomalies or emergencies. In collaboration with the SAP, the OCC ensures that the wind farm's operation remains safe and efficient.

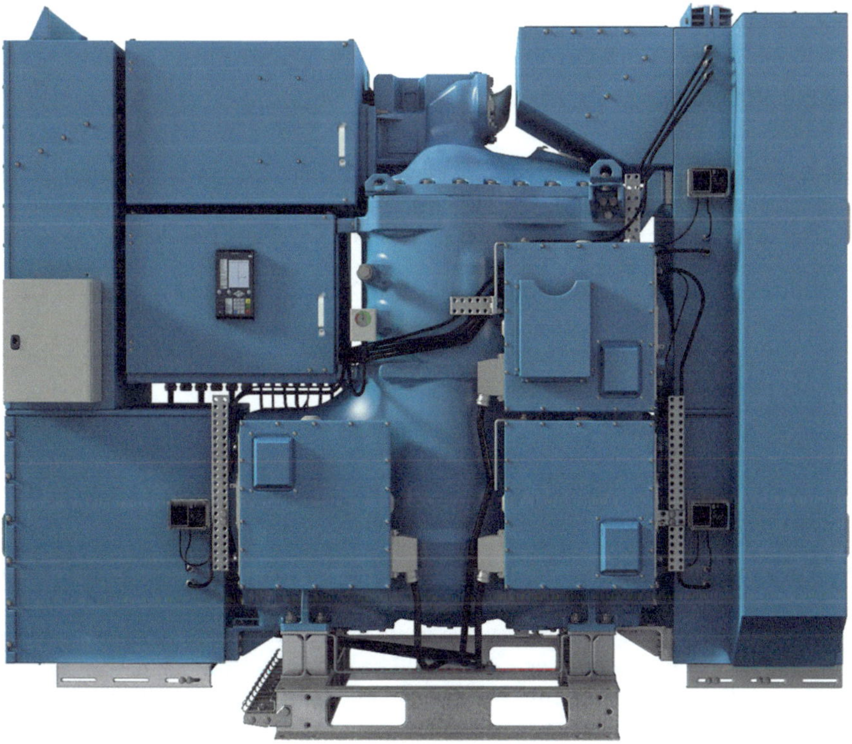

Fig. 8.3 8VM1 switchgear, https://www.siemens.com/global/en/products/energy/topics/sustainability-for-electrification-and-automation.html, Nov. 2024—Copyright by Siemens Energy

8.4.4 *Switching Order*

The Switching Order is the central document for each switching operation and contains all necessary information, such as:

- The exact location and affected switching devices.
- The sequence of switching actions.
- Safety measures, such as locking and grounding installations.
- Responsibilities and contact information of participants.

The Switching Order is prepared and reviewed by the SAP before being confirmed by the OCC and the AP. Upon completion of the switching operation, it is converted into a switching protocol that serves as documentation.

8.5 Practical Execution of a Switching Operation

8.5.1 *Preparation by the AP*

On-site, the AP begins by reviewing the Switching Order and necessary safety measures. All required tools, key systems, and protective devices are checked for availability and condition. Coordination with the OCC ensures that the switching operation can commence.

8.5.2 *Switching Communications*

A central element of the switching operation is the switching conversation conducted between the AP and the OCC or another qualified person on-site. Communication follows a clearly defined pattern to avoid misunderstandings:

- **Announcement**: The AP states the planned action, e.g., "Opening circuit breaker—QZxx."
- **Confirmation**: The OCC repeats the instruction and provides confirmation, e.g., "Understood, circuit breaker—QZxx will be opened—affirmative."
- **Completion**: After performing the switching action, the AP reports the status, e.g., "Circuit breaker—QZxx opened."

If the OCC raises concerns, for example, due to parallel work in adjacent areas, the switching operation is immediately halted until the situation is clarified.

8.5.3 *Execution of Switching Actions*

The AP performs the switching actions strictly according to the Switching Order. Each action is logged, including timestamps and status. Special attention is paid to internal interlocks of the switching devices to ensure that no dangerous conditions arise, such as switching onto a grounded line.

8.5.4 *Completion and Handover*

After completing all steps, the AP informs the OCC about the system's status and passes the switching protocol to the SAP. The SAP reviews the protocol and confirms the safe execution.

8.6 Challenges and Solutions

8.6.1 Complex Switching Operations

Especially during the initial energization of new installations, the multitude of necessary steps and dependencies can be challenging. A clearly structured Switching Order and regular switching communications minimize the risk of errors.

8.6.2 Communication

Error-free communication between the AP, SAP, and OCC is essential. Standardized terminology and clear instructions are crucial.

8.6.3 Safety Measures

All safety precautions must be strictly observed. This includes locking and grounding installations and adherence to Lockout-Tagout (LOTO) processes.

8.6.4 Sanction for Testing: An Integral Part of Switching Operations

The topic of **Sanction for Testing (SFT)** is an indispensable part of the safe execution of switching operations, especially during the commissioning of new offshore wind farm components. This document (Fig. 8.4) allows controlled tests on electrical installations that are typically outside standard operational releases. Due to its central importance for safety and efficiency, the concept of SFT is revisited in this chapter to highlight its connection with switching operations and the responsibilities of various participants.

8.6.5 Purpose of the Sanction for Testing

An SFT serves to separate specific electrical installations or parts thereof from their regular operational release for testing purposes. This ensures that these areas can be safely worked on or tested without affecting the normal operation of other systems. SFT is particularly relevant during initial energization, as extensive tests are often required to confirm the functionality and safety of the overall system.

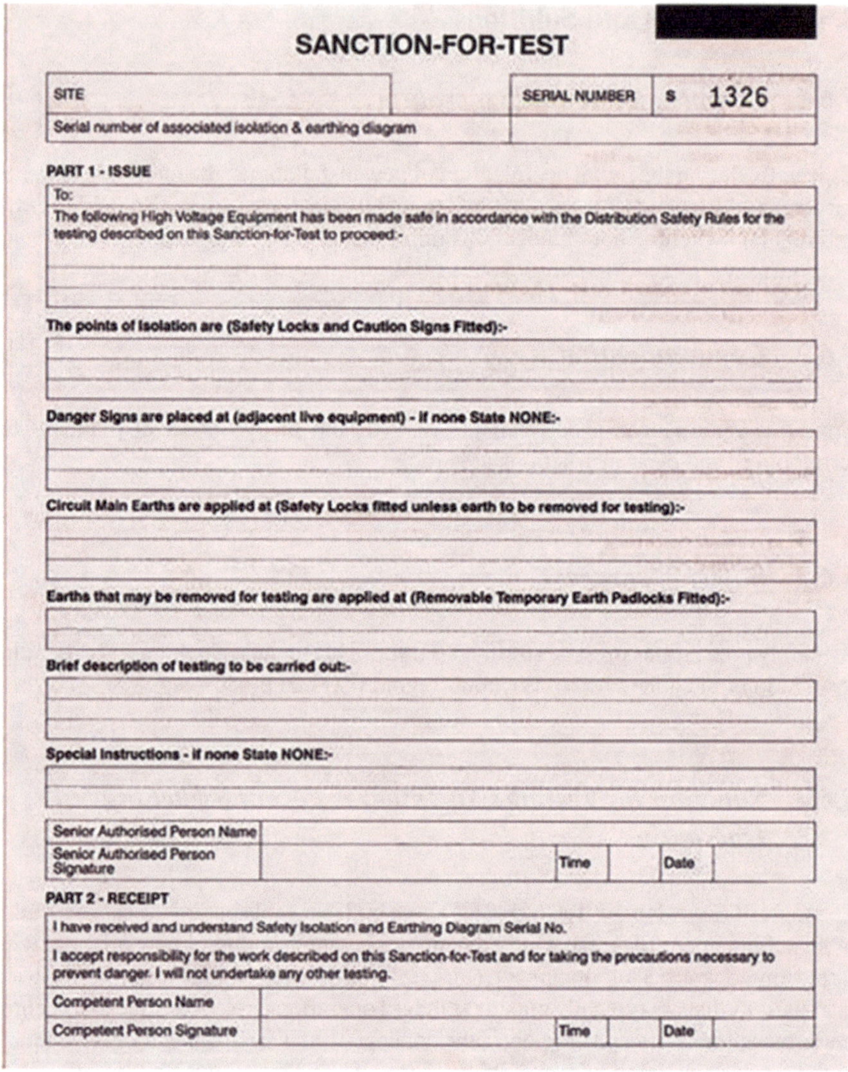

SANCTION-FOR-TEST

SITE	SERIAL NUMBER	S 1326
Serial number of associated isolation & earthing diagram		

PART 1 - ISSUE

To:

The following High Voltage Equipment has been made safe in accordance with the Distribution Safety Rules for the testing described on this Sanction-for-Test to proceed:-

The points of Isolation are (Safety Locks and Caution Signs Fitted):-

Danger Signs are placed at (adjacent live equipment) - If none State NONE:-

Circuit Main Earths are applied at (Safety Locks fitted unless earth to be removed for testing):-

Earths that may be removed for testing are applied at (Removable Temporary Earth Padlocks Fitted):-

Brief description of testing to be carried out:-

Special Instructions - If none State NONE:-

Senior Authorised Person Name					
Senior Authorised Person Signature		Time		Date	

PART 2 - RECEIPT

I have received and understand Safety Isolation and Earthing Diagram Serial No.

I accept responsibility for the work described on this Sanction-for-Test and for taking the precautions necessary to prevent danger. I will not undertake any other testing.

Competent Person Name					
Competent Person Signature		Time		Date	

Fig. 8.4 Sanction for Test, old but still used, https://powerdistributionqa.wordpress.com/safety-documents/, Nov. 2024

8.6.6 Connection with Switching Operations

The use of an SFT is directly related to switching operations:

- **Isolation and Separation**: Before a test begins, the relevant installations must be safely isolated and disconnected through a switching operation. This is carried out by the AP and supervised by the SAP.

- **Creation of a Switching Order**: The Switching Order includes the exact steps to implement the SFT, including isolation, grounding, and handing over the isolated area to the test team.
- **Communication with the OCC**: The OCC is informed about the SFT and associated switching operations to ensure that no unintended interventions or conflicts with other work occur.

8.6.7 Responsibilities in the SFT

Proper implementation of an SFT requires clear responsibilities:

- **Senior Authorized Person (SAP)**: The SAP approves the SFT and ensures that all involved parties are informed about the testing measures. They verify that all switching operations are completed and safely executed before tests begin.
- **Authorized Person (AP)**: The AP performs the necessary switching measures and documents them according to the SFT requirements.
- **Test Team**: The team conducting the tests assumes responsibility for complying with the specific provisions of the SFT and reports the status of the tested installations back to the SAP upon completion.

8.6.8 Example: Use of SFT During Initial Energization

Suppose a protection relay needs to be tested during the initial energization of an offshore wind farm. The SFT specifies that the affected area is safely isolated so that the test team can examine the relay's function without risk. A Switching Order describes the isolation and grounding of the corresponding switchgear. After completing the tests, the installation is returned to normal operational status, and the SFT is withdrawn.

Note: The accurate and safe execution of energization and switching operations is critical for the successful commissioning of offshore wind farms. Adhering to standards, clear communication, and understanding the roles and responsibilities of all involved parties ensure that these complex processes are carried out efficiently and safely.

Chapter 9
Management of Temporary Power Supply Systems

9.1 The Necessity of Temporary Power Supply Systems

During the construction and commissioning phases of an offshore wind farm, temporary power supply (TPS) plays a crucial role. It bridges the period when the regular grid connection is not yet available and ensures that essential activities—such as the installation of wind turbine generators (WTGs), the commissioning of cable systems, and the integration of the substation—proceed smoothly. TPS is thus an indispensable element in the construction organization of offshore wind projects.

9.1.1 WTG Installation as a Driving Force

The primary driver for the use of TPS is the installation of the wind turbine generators. Offshore wind farms often consist of dozens to hundreds of turbines erected in strict sequence. These installations require power for cranes, tools, control systems, and lighting. Additionally, the internal systems of the WTGs must be tested before the final grid connection, which also demands a reliable energy source.

9.1.2 Coordination with the Project Schedule

The implementation of TPS is closely linked with the planning of the grid connection, the substation, and the cable commissioning. Delays in any of these areas can prevent the WTG installation from proceeding as planned. Therefore, precise coordination among the involved trades is essential. The project schedule must be flexible enough

C. Hoppe, *Commissioning Management in Offshore Wind Projects*,
SpringerBriefs in Applied Sciences and Technology,
https://doi.org/10.1007/978-3-032-05358-9_9

to account for unforeseen events, such as adverse weather conditions or logistical bottlenecks. In such situations, TPS acts as a safety net to maintain progress and prevent jeopardizing the installation.

9.2 The Timing Factor and Mobilization of TPS

Setting up a temporary power supply system requires careful planning. The need for TPS often begins parallel to the initial installations of the substation and cabling and increases as the first WTGs are installed. Ramp-up phases are essential here, as both the energy requirements and the complexity of the network increase with the completion of construction work. Mobilization costs and times must be integrated into the project plan early to avoid delays.

9.2.1 Experience and Reliability Through Full-Service Providers

A decisive aspect in selecting a TPS system is the reliability of the power supply. Offshore locations impose special requirements on stability and operational capability, as an interruption not only causes delays but also poses significant safety risks.

9.2.2 The Role of Full-Service Companies

Industry experience shows that full-service providers (Fig. 9.1) for diesel generators are reliable partners. These companies not only supply the necessary equipment but also handle maintenance, servicing, and fuel management. They guarantee high availability of the generators, which is particularly crucial in offshore projects where access to installations is often limited by weather conditions.

9.2.3 Maintenance and Servicing as Keys to Stability

Regular maintenance intervals are essential to ensure the operational readiness of the generators. In offshore environments, this requires specially trained personnel capable of working under challenging conditions. Maintenance encompasses both mechanical and electrical components to minimize the risk of failures.

Fig. 9.1 TPS, Manor Renewable Energy, 2022—Copyright by MRE

9.2.4 Fueling Strategies During Adverse Weather Periods

A robust fueling management plan is vital to ensure power supply during periods of bad weather. This includes planning fuel reserves and organizing tank vehicles or ships that can operate even under adverse conditions. A detailed fueling schedule reduces dependence on spontaneous resupply deliveries and ensures an uninterrupted power supply.

9.2.5 Minimizing the Use of TPS

While TPS is indispensable, the goal of every project should be to reduce the use of temporary power supply systems as much as possible. A well-planned and executed grid connection can minimize dependence on TPS, but it would be unwise to exclude TPS entirely from project planning.

9.2.6 Demand Planning as a Foundation

Precise demand planning is crucial for the effective utilization of TPS. This involves determining the required amount of energy, selecting suitable generators, and planning personnel deployment. Early integration of these factors into the project plan can significantly reduce costs and mobilization times.

9.2.7 Lead Times and Ramp-Up Phases

Establishing TPS requires extensive preparations, especially when transporting generators, fuel, and personnel to the site. These lead times must be accounted for in the project plan to avoid bottlenecks. Ramp-up phases, during which energy demand gradually increases, offer opportunities to deploy TPS more efficiently.

9.2.8 Minimization During Adverse Weather Periods

During expected bad weather periods, such as autumn or winter, TPS should ideally no longer be necessary. This assumes that the grid connection and permanent power supply are operational by that time. Planning should aim to bridge these critical periods with minimal TPS dependency.

9.3 Alternatives to Diesel Generators: A Critical Review

In light of the environmental impact of diesel generators, alternative solutions for temporary power supply in offshore wind farms are increasingly being explored. These approaches aim to reduce emissions and lower operating costs. However, they also introduce new challenges.

9.3.1 *Successful Projects and New Technologies*

Some offshore projects have successfully tested alternative TPS solutions. For example, a company provided an adapted temporary power supply for an offshore wind farm in the German Bight to protect a significant investment [1].

Additionally, environmentally friendly emergency generators have been installed on the converter platform of the Borkum Riffgrund 3 offshore wind farm, providing an eco-friendly emergency power supply with IMO Tier III-certified units [2].

9.3.2 *The Self-sustaining Turbine System by Siemens Gamesa*

A noteworthy approach is the Self-Sustaining Turbine (SST) system developed by Siemens Gamesa. This system utilizes the idle mode of wind turbines to generate power for their own consumption, reducing the need for external energy sources during commissioning. However, implementing such systems requires careful planning and adaptation to specific project requirements [3].

9.3.3 *Challenges of Alternative Solutions*

While alternative TPS technologies are promising, they often place high demands on planning, integration, and maintenance. Their reliability in offshore environments needs further validation before they can serve as full-fledged replacements for diesel generators. Additionally, the investment costs for such systems are often higher, and their efficiency may depend on environmental conditions.

9.4 Conclusion

Alternative solutions for temporary power supply in offshore wind farms offer significant potential for reducing emissions and operating costs. Nevertheless, they require careful evaluation and planning to ensure they meet the specific requirements and challenges of the offshore environment. While TPS remains a critical component during the construction and commissioning phases, the strategic integration of reliable alternatives can enhance sustainability and operational efficiency in future projects.

References

1. Aggreko Ltd., *Modifizierte Generatoren versorgen Offshore-Windpark* (2024). Von https://www.aggreko.com/de-de/case-studies/renewables/generators-help-power-wind-farm abgerufen
2. MTU, L.M., *Umweltfreundlicher Notstrom auf hoher* (2021). See Von https://www.mtu-solutions.com/eu/de/stories/energie/gas-generatoren/green-emergency-power-on-the-high-seas.html abgerufen
3. Siemens Gamesa Renewable Energy, S.A.U., *The Future for Offshore wind Power Looks Brighter Than Ever* (2024). Von https://www.siemensgamesa.com/global/en/home/products-and-services/offshore.html abgerufen

Part IV
Coordination, Management, and Handover

The successful implementation of an offshore wind project requires precise coordination and smooth handover between the various project phases. Part IV is dedicated to the central management tasks during commissioning and the critical transition from the construction phase to the operational phase. The focus is not only on practical and organizational challenges but also on the processes and interfaces that enable efficient communication and collaboration among all parties involved.

The chapters in this part illuminate both the strategic and operational aspects of management. They cover topics such as the establishment of an effective reporting system, the organization of meetings, communication between onshore and offshore teams, and the role of central coordination points like the Operation Control Center (OCC) and the Marine Coordination Center (MCC). Additionally, the handover processes essential for the successful start of the operational phase are discussed.

This part offers a comprehensive perspective on the management tasks in offshore wind projects and demonstrates how clearly defined processes, close collaboration, and structured communication lay the foundation for sustainable and efficient project execution.

Chapter 10
Handover and Takeover Processes

The handover processes play a central role in the lifecycle of offshore wind farms. A clear understanding of the differences between "handover" and "takeover," as well as the implementation of effective tools and procedures, is crucial for a smooth transition from the construction to the operational phase.

10.1 Difference Between Handover and Takeover

The term **"handover"** refers to the formal transfer of responsibility, control, or ownership from one party to another. In the context of offshore wind farms, this means transferring the facility from the construction or project team to the operations and maintenance (O&M) team. **"Takeover,"** on the other hand, describes the process by which the receiving party assumes responsibility and control. While the handover process initiates the transfer, the takeover process marks the completion of this transfer and the beginning of the operational phase.

10.2 Creation of Checklists and Handover Protocols

The creation and use of checklists and handover protocols are fundamental components of the handover and takeover process. These documents not only serve to systematically record inspection and acceptance criteria but also ensure that all relevant requirements—both contractual and technical—are met. Their precise drafting and utilization are essential to make the handover process transparent and efficient.

C. Hoppe, *Commissioning Management in Offshore Wind Projects*,
SpringerBriefs in Applied Sciences and Technology,
https://doi.org/10.1007/978-3-032-05358-9_10

10.2.1 Consideration of Contract Contents

Checklists and handover protocols must be closely aligned with the defined contract contents. The standards, requirements, and conditions stipulated in contracts form the basis for the points to be inspected. This includes acceptance conditions, warranty agreements, and performance descriptions. It is crucial that the checklists are structured to ensure complete and thorough fulfillment of these requirements. Each inspection step should be clearly assignable to a specific contract section.

10.2.2 Technical Requirements and Specifications

The technical requirements defined in the specifications and requirements documents must be integrated in detail into the checklists. This involves not only verifying the technical fulfillment of the requirements but also ensuring that they can be reliably maintained under real operating conditions. For example, parameters such as the thermal load capacity of electrical components or the correct calibration of monitoring and control systems could serve as central inspection points.

10.2.3 Interfaces to Subsequent Trades

An important aspect in creating handover protocols is considering the interfaces to subsequent trades. These interfaces must be clearly documented and evaluated, as defects or uncertainties in one trade can directly impact subsequent work. It must be ensured that handover points like cable connections, communication protocols, or mechanical linkages are fully tested and accepted before the next trade begins its work.

10.2.4 Clarity of Inspection Points

A common problem in practice is the ambiguity of inspection points, which can lead to misunderstandings and delays. To avoid this, checklists and protocols should be designed so that each inspection step is clearly and unambiguously formulated. A description like "Cable test completed" is insufficient; instead, specific details such as the type of test performed (e.g., insulation measurement), the standards applied, and the accepted threshold values should be included. Additionally, inspection points must be defined so they are interpreted consistently, regardless of who performs them.

10.2.5 Consideration of Local and Technical Dependencies

Another crucial point is considering dependencies within an inspection protocol. Many systems and components are technically and spatially interdependent, requiring a logical sequence of inspections. For example, inspecting a cable protection system can only occur after the cable itself has been tested and properly installed. These dependencies must be clearly defined and accounted for in the checklists and protocols to ensure a structured process.

10.2.6 Documentation and References

The availability and reference to relevant documentation play a key role. Handover protocols should not only document the inspections conducted but also refer to associated reports, certificates, and technical documents. This facilitates not only the traceability of inspections but also allows quick access to relevant information for future questions or issues.

10.2.7 Escalation Mechanisms in Case of Safety Risks

For safety–critical equipment such as fire alarm systems, gas extinguishing systems, or high-voltage components, escalation mechanisms in case of deviations are essential. These should be clearly defined in the protocols and include a hierarchical sequence of actions. For example, an escalation could involve immediate reporting to the HSE Manager, a technical re-inspection by experts, or the suspension of all subsequent work.

10.2.8 Assessment and Prioritization of Deviations

Not all deviations have the same impact on the project's progress. Therefore, it's important to implement mechanisms for assessing and prioritizing deviations in the handover protocols. These should clearly distinguish between critical defects that must be remedied immediately and less urgent deviations that can be addressed within a defined timeframe. This enables efficient resource planning and minimizes delays.

10.2.9 Additional Considerations

- **Digitization of Processes**: Utilizing modern software solutions like iAuditor (Fig. 10.1) can significantly simplify the creation and management of checklists and protocols. They allow for real-time documentation of inspections, automatic report generation, and centralized storage.
- **Training of Personnel**: Regular training of involved personnel is essential to ensure that checklists are correctly filled out and interpreted.
- **Traceability and Auditability**: All protocols should be designed to be easily reviewable during external audits.

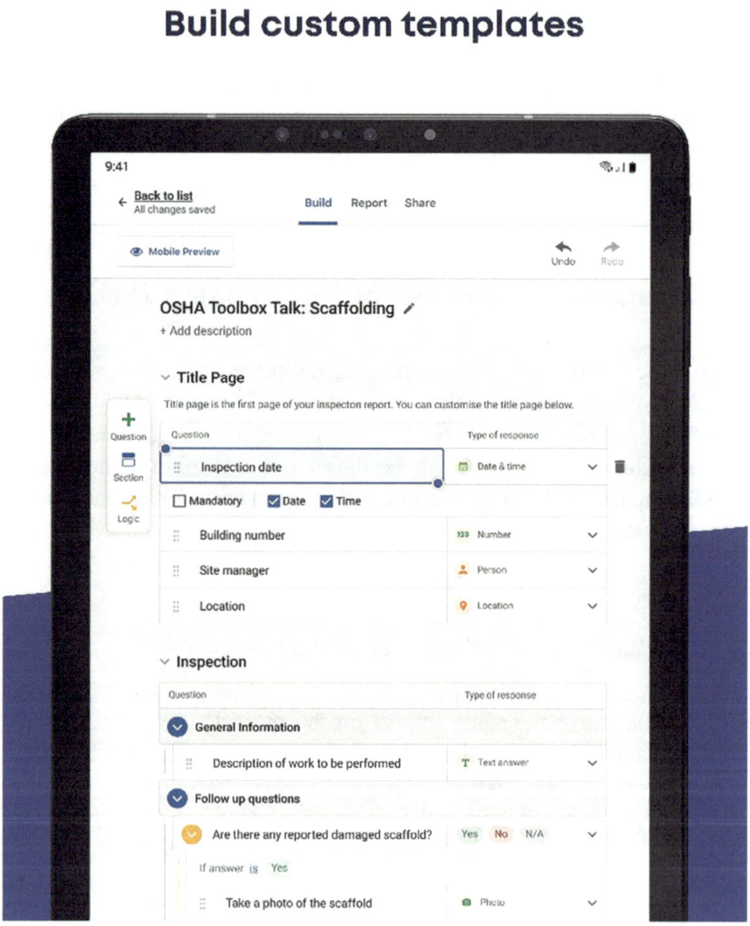

Fig. 10.1 iAuditor, template creation, https://safetyculture.com/checklists/construction/. Nov. 2024—Copyright by SafetyCulture

Checklists and handover protocols are more than just administrative documents—they are indispensable tools for ensuring quality and safety in the handover process. Through precise formulations, consideration of interfaces and dependencies, and the incorporation of modern technologies, a smooth and efficient handover process can be guaranteed. They lay the foundation for a transparent and traceable transfer that protects all parties equally. Errors, damages, or incompleteness can be recorded and attributed to suppliers or contractors. This tracking of Non-Conformities (NCs) is of significant importance for contract and claim management, which will be discussed in more detail in Part 5, Chap. 14.

10.3 Integration with Operations & Maintenance (O&M)

The collaboration between Commissioning Management and Operations & Maintenance (O&M) is one of the central prerequisites for the successful transition of an offshore wind farm from the construction to the operational phase. This close connection ensures that all technical, organizational, and documentation requirements are considered during both the handover and takeover processes. Only through seamless integration of these two areas can the long-term operational safety and efficiency of the facility be ensured.

10.3.1 Joint Development of Checklists and Protocols

An effective handover process requires checklists and handover protocols precisely tailored to the O&M team's needs. Experience shows that it's particularly valuable to develop these documents jointly with the O&M team, if the project situation allows. This way, specific operational requirements—such as regular maintenance intervals, availability of spare parts, or special inspection points—can be considered from the outset.

This close collaboration ensures that all relevant aspects are included in the handover documentation, preventing future misunderstandings. Simultaneously, the O&M team can provide feedback during the commissioning phase, allowing for timely optimizations or adjustments.

10.3.2 Assumption of Warranty Claims

A crucial point in the handover process is the assumption of warranty claims by the O&M team. During the handover phase, it's precisely documented which systems, components, and services are covered under warranty and which are not. This specifically involves:

- **Manufacturer Warranties**: The warranty obligations of turbine manufacturers and key component suppliers must be clearly defined and documented.
- **Punch Lists and Defects**: Open items that couldn't be fully resolved during commissioning must be explicitly listed in the handover protocol. This allows the O&M team to track these points or assert claims.
- **Safety–Critical Defects**: Special attention is given to deviations affecting the facility's safety. These must be prioritized and either remedied before handover or addressed through clear escalation measures.

A thorough handover documentation encompassing all relevant warranty claims and responsibilities is therefore indispensable. These documents serve not only for traceability but also as a basis for potential future disputes or claims.

10.3.3 Documentation of the Wind Farm

Complete and structured documentation of an offshore wind farm is essential not only during the handover process but also for long-term operation. The O&M team requires access to all relevant information to operate and maintain the facility efficiently. This documentation includes:

- **Technical Documents**: Circuit diagrams, cable route plans, operating manuals, and calibration reports.
- **Inspection and Test Protocols**: Results from inspections and tests conducted during commissioning, including load tests and switching protocols.
- **Operation and Maintenance Manuals**: Comprehensive guides covering all operational and maintenance aspects, tailored to the specific requirements of the wind farm.
- **Digitized Systems**: Many projects today utilize digital platforms to centrally manage all documents and provide real-time access to the O&M team.

The quality of the handover documentation directly impacts operational efficiency and safety. Incomplete or poorly structured documentation can lead to delays, increased maintenance efforts, and, in the worst case, safety risks.

10.3.4 Close Collaboration During the Transition Phase

The transition phase between commissioning and O&M is a critical period requiring regular exchange between both teams. Key points include:

- **Training the O&M Team**: The commissioning team should ensure that the O&M team is fully familiar with the facility's specific requirements and characteristics. This includes both technical training and practical on-site orientations.

- **Transfer of Responsibilities**: A clearly defined schedule for transferring responsibilities ensures that the O&M team is gradually introduced to managing the facility without any supervision gaps.
- **Feedback Loops**: Regular feedback sessions during the transition allow for early clarification of open issues and continuous improvement of handover processes.

10.4 Conclusion

Integrating O&M into the handover and takeover process is essential for ensuring the long-term success of an offshore wind farm. Through close collaboration in creating checklists and protocols, clear definition of warranty claims, and providing comprehensive documentation, a smooth transition can be secured. This not only enhances operational efficiency but also establishes the foundation for sustainable and safe utilization of the facility.

Chapter 11
Communication and Reporting

11.1 Meetings

Meetings are an indispensable part of commissioning management and form the backbone of effective communication within offshore wind projects. Through structured and targeted discussions, information can be exchanged efficiently, problems can be identified promptly, and solutions can be developed collaboratively. In commissioning management, various meeting formats are utilized, specifically tailored to the requirements of this project stage.

11.1.1 Fundamentals of Effective Meetings

A central pillar of successful meetings is adherence to general principles that are firmly established in the common doctrine of meetings. Each meeting should start with a clear objective known to the participants in advance. A well-prepared agenda helps maintain focus and efficiently address all relevant topics. Time management is a decisive factor, as adhering to the scheduled duration not only improves the meeting's flow but also protects the productivity of the participants. Active participation is encouraged to bring in as many perspectives as possible, while discussions should always be outcome-oriented. At the end of each meeting, clear decisions and defined next steps make the results tangible.

To document the outcomes of the meeting and make them available to participants and absentees alike, minutes in the form of **Minutes of Meeting (MoM)** are created. These minutes contain the most important discussions, decisions made, and established next steps. They ensure traceability and serve as a binding reference for all participants and stakeholders. MoMs should be created immediately after the meeting and distributed to all relevant parties to establish a unified information base.

C. Hoppe, *Commissioning Management in Offshore Wind Projects*,
SpringerBriefs in Applied Sciences and Technology,
https://doi.org/10.1007/978-3-032-05358-9_11

11.1.2 Daily Calls: Daily Coordination

Daily coordination meetings, known as **Daily Calls**, serve to capture the current status of the project and coordinate daily tasks. These short meetings, typically lasting 15–30 min, focus on the previous day's progress, identified problems, and planned activities for the current day. They enable the team to react quickly to new developments and ensure that all involved are on the same page.

Especially in the dynamic environment of an offshore wind project, Daily Calls are an effective means of staying on top of things and recognizing potential challenges early. Their short duration and focused agenda ensure that they do not disrupt the workflow but rather support it.

11.1.3 Weekly Update Meetings: The Weekly Overview

Complementing daily coordination, **Weekly Update Meetings** provide a more comprehensive overview of project progress. These meetings, attended by project team members, client representatives, and relevant stakeholders, focus on milestones achieved during the past week, review of the project schedule, and planning for the coming week. With a typical duration of 60–90 min, they provide sufficient space to analyze problems in detail and develop solutions.

A crucial aspect of these meetings is the creation of a comprehensive protocol that documents all discussed topics, results, and established actions. These MoMs ensure that all participants have a unified information base and that decisions made can be traced transparently. They also serve as a basis for planning the next week and as a reference for future discussions.

11.1.4 Task Force Meetings: Focused Problem Solving

For specific topics or urgent issues, **Task Force Meetings** are convened. These meetings are thematically focused and bring together specialists to develop targeted solutions or address special tasks. In such meetings, the problem or task is first clearly defined before analyzing the causes and developing solutions. Responsibilities are clearly assigned, and deadlines are set for implementing the agreed-upon measures.

Task Force Meetings offer the advantage of concentrating on a specific problem and are therefore particularly effective in handling complex or urgent topics. Their duration varies depending on the complexity of the issue, but the focus is always on quick and targeted solution development.

11.1.5 Monthly Project Report Meetings: Strategic Planning

Another important format in commissioning management is the **Monthly Project Report Meetings**. These more extensive meetings provide a detailed insight into the project status and serve to evaluate performance, strategic planning, and decision-making at the leadership level. Besides presenting project progress and analyzing budgets and financial aspects, these meetings also assess risks and opportunities to make necessary strategic adjustments.

These sessions, which often take several hours, create transparency and ensure that all decision-makers are on the same page. The protocols developed in these meetings are of crucial importance for traceability and serve as a foundation for strategic decisions in the further course of the project.

11.1.6 Overview

By applying these different meeting formats, a structured exchange of information is ensured, which significantly supports project success. Each meeting format fulfills a specific function, whether it's daily coordination, weekly planning, or strategic alignment. Together, they form an effective communication network that ensures all involved are on the same page, problems are identified in a timely manner, and the project progresses purposefully. Protocols like the Minutes of Meeting are a decisive tool for recording results and promoting transparency within the project.

11.2 Establishing an Effective Reporting System

A structured and well-organized reporting system is essential for the success of an offshore wind project, particularly during the sensitive phase of commissioning management. Reports provide transparency, create a clear information base, and serve as decision-making tools at all levels of project management. Three central report types—Daily Progress Reports, Weekly Progress Reports, and 3-Day Look-Ahead Reports—play a crucial role by covering different perspectives and time frames.

11.2.1 Daily Progress Reports: Documenting Daily Progress

Daily Progress Reports (DPRs) are an indispensable tool for documenting the project's daily progress and ensuring that all involved are up to date. These reports provide a snapshot of project activities and focus on events of the previous day as well as plans for the current day. They include an overview of completed tasks, progress in ongoing work, and any challenges or delays encountered. Safety-related incidents, such as near-misses or unforeseen events, are also documented. Besides this information, weather conditions often play a central role, as they can have significant impacts on work in offshore environments.

An important component of a DPR is short-term planning, offering a preview of the day's activities. Necessary adjustments to the original plans are considered. Daily Progress Reports are not only a basis for Daily Calls but also promote quick decisions and enable responsive project control. Their conciseness and daily relevance make them a vital element of reporting.

11.2.2 Weekly Progress Reports: Weekly Progress Overview

Weekly Progress Reports (WPRs) offer a more comprehensive perspective and are designed to present the project's progress over a week. They serve to document achieved milestones, identify risks, and plan strategically for the coming week. Unlike DPRs, which focus on operational details, a WPR focuses on a strategic summary that can be used both internally and externally. A WPR typically contains a detailed overview of the project status, including an assessment of time and budget progress compared to the original plans. It also provides insights into resource utilization, such as personnel, materials, and equipment, and evaluates their efficiency.

Another central point is risk analysis, which identifies new potential risks and shows progress in addressing existing risks. Deviations are documented, and their impacts on the schedule or budget are assessed. Finally, a WPR often offers recommendations for action and escalations that form a basis for decisions by project management and stakeholders. The structured and regular creation of WPRs ensures strategic orientation and keeps the project on course.

11.2.3 3-Day Look-Ahead Reports: Short-Term Planning

The **3-Day Look-Ahead Report (3DLAR)** is specifically tailored to the short-term planning of the upcoming three days. It combines elements of progress documentation with a detailed preview and serves as an operational planning tool that bridges Daily Progress Reports and Weekly Progress Reports. The report begins with a summary of ongoing work and indicates which tasks are planned for the next three days. Special attention is given to the required resources to ensure that materials, personnel, and equipment are available on time.

A 3DLAR also highlights dependencies that exist between different tasks or teams and identifies potential challenges that need to be resolved in the short term. These reports are particularly useful for reacting flexibly to changes and avoiding delays. Through their precise and current orientation, 3DLARs are an indispensable tool for daily and weekly planning.

11.2.4 Integration of Report Types

An effective reporting system relies on integrating these report types into a unified system. Daily Progress Reports provide current data and promote quick responses to operational challenges. Weekly Progress Reports create a strategic overview and ensure that long-term goals are met. The 3-Day Look-Ahead Report bridges these two formats and ensures precise and flexible planning. Together, these reports form an effective network that creates transparency, enables informed decisions, and sustainably supports project success.

The use of modern software solutions, such as iAuditor or specific project management tools, can significantly simplify the creation and distribution of these reports. They enable real-time documentation, automatic distribution, and centralized data storage, making reporting more efficient and less prone to errors. Such a system ensures that reporting not only meets current requirements but is also flexible enough to address future challenges.

11.3 Communication Between Onshore and Offshore Teams

Communication between onshore and offshore teams is one of the biggest challenges in commissioning management of offshore wind projects. Due to the physical separation and often challenging working conditions at sea, this collaboration requires careful organizational planning and a clear management structure. The following focuses not on technical solutions like communication systems or data transmission paths but on the organizational prerequisites and management tasks necessary to ensure effective and smooth communication.

11.3.1 Establishing Uniform Communication Structures

One of the central management tasks is establishing uniform communication structures that define clear roles, responsibilities, and procedures. It must be precisely determined what information is exchanged between the onshore and offshore teams, who is responsible for transmitting it, and how feedback is provided. A hierarchically organized communication structure can help avoid misunderstandings and ensure that the right information reaches the right people.

A central contact person, often in the role of an Offshore Site Manager or Coordinator, serves as the interface between the two teams. This person ensures that information is consistently passed on and acts as a bridge between the offshore and onshore teams.

11.3.2 Clear Guidelines for Meetings and Reports

Meetings and reports are essential elements of communication between onshore and offshore. Regular meetings—such as daily calls or weekly updates—enable continuous exchange of status reports, challenges, and planned measures. These meetings should have a fixed agenda and clear objectives to be effective.

Moreover, reports are an important part of communication. Offshore teams must create daily progress reports that are consolidated and analyzed onshore. These reports should be standardized to ensure consistency and comparability. Onshore teams, in turn, must ensure that relevant information—such as planned switching operations or resource availabilities—is communicated to offshore teams in a timely manner.

11.3.3 Training and Awareness

Effective communication requires a deep understanding of the working conditions and challenges of the respective teams. Onshore managers need to understand the specific requirements and limitations of offshore teams, while offshore personnel should consider the strategic and planning constraints of onshore teams. Regular training and workshops can help promote this understanding.

Another important aspect is cultural awareness. Offshore projects are often international, and different cultures can bring different communication styles. Management should be aware of these differences and offer training to promote a unified communication culture.

11.3.4 Escalation Processes and Conflict Management

Despite all organizational efforts, communication problems or conflicts can occur in a complex project like an offshore wind farm. Escalation processes are therefore essential to manage such situations. Management must clearly define how and when problems are escalated and who is responsible for resolving them. These processes should be transparent and easily understandable for all involved.

Equally important is proactive conflict management. Through regular feedback sessions and promoting an open communication culture, potential conflicts can be recognized and resolved early before they escalate.

11.3.5 Communication Plans and Documentation

A comprehensive communication plan is an indispensable tool to manage communication between onshore and offshore teams. This plan should clearly define communication channels, the frequency of meetings and reports, and the responsibilities of individual team members. It serves as a guide for all involved and ensures that no important information is lost.

Documentation also plays a central role. All important decisions, agreements, and reports must be systematically documented to ensure seamless traceability. This is important not only for current project progress but also for future audits or lessons learned.

11.4 Conclusion

Communication between onshore and offshore teams requires careful organizational planning and a clear management structure. Uniform communication structures, regular meetings and reports, targeted training, and clear escalation processes form the foundation for effective collaboration. A well-thought-out communication plan and systematic documentation ensure that collaboration functions smoothly and that teams act as a unit despite physical separation. By fulfilling these organizational prerequisites, project management can ensure that the challenges of offshore communication are successfully overcome.

11.5 Role of the OCC as a Central Communication Point

The **Operation Control Center (OCC)** is one of the most important organizational units in offshore wind projects and assumes the function of a central communication point. It serves as a hub for exchanging information among various project participants, including onshore and offshore teams, commissioning management, O&M teams, and external stakeholders. Through its central role, the OCC significantly contributes to coordination, monitoring, and safety throughout the commissioning phase.

Fig. 11.1 Control Center, Pyrotech Workspace, https://www.pyrotechworkspace.com/blog/customer-appreciation, Nov. 2024

11.5.1 Overview of the OCC's Tasks

The OCC is primarily responsible for monitoring and controlling all electrical and mechanical systems of an offshore wind farm. Core tasks include real-time monitoring of operating parameters, coordination of switching operations, and support in troubleshooting. Simultaneously, the OCC assumes a key function in the communication chain by acting as an intermediary among the involved parties and ensuring the information flow (Fig. 11.1).

One of the most important functions of the OCC is logging and documenting all operations. This includes recording switching operations, communication between onshore and offshore teams, and documenting disturbances and their resolution. This data is essential to ensure transparency and identify potential problems early.

11.5.2 Coordination and Approval of Switching Operations

A central task of the OCC is coordinating and approving switching operations. The OCC acts as a control authority that ensures all necessary safety measures have been taken before a facility or system is switched. The Senior Authorized Person (SAP) and the Authorized Person (AP) work closely with the OCC to ensure switching operations are conducted correctly and safely.

The OCC reviews the switching instructions, also known as Switching Orders, for completeness and correctness. It also ensures that all involved parties—including offshore teams and external service providers—are informed about the planned switching operation. Through this coordination, the OCC minimizes the risk of errors and accidents that could arise from communication gaps.

11.5.3 Communication with Offshore Teams

Communication with offshore teams is another essential task of the OCC. Due to the physical distance and often limited communication possibilities at sea, the OCC is frequently the only communication channel between offshore teams and other project participants. It undertakes the task of transmitting information quickly and precisely to ensure offshore personnel are informed about all relevant developments.

An example of this role is support during emergencies or unexpected disturbances. In such cases, the OCC coordinates the necessary measures, establishes communication among the involved parties, and ensures the situation is resolved quickly and efficiently.

11.5.4 Support of Commissioning Management

The OCC is also an indispensable support for commissioning management. It ensures that all data collected during the commissioning phase—such as switching protocols, test reports, and operating parameters—are centrally recorded and evaluated. This information is crucial for assessing project progress and identifying potential problems.

Furthermore, the OCC serves as a central contact point for coordinating resources, such as personnel and equipment needed for specific tasks offshore. Through this central role, the OCC significantly contributes to making processes efficient and smooth.

11.5.5 Safety Monitoring and Escalation Management

Another central aspect of the OCC's work is safety monitoring. It continuously monitors operating conditions and promptly reports deviations or potential hazards to the appropriate parties. In the event of a safety-related incident, the OCC takes over the coordination of escalation and ensures that all involved are informed and necessary measures are taken.

The OCC's escalation management includes involving HSE managers, technical leaders, and external experts when required. Through this function, the OCC contributes to resolving safety issues quickly and effectively.

11.6 Role of the MCC as a Communication Point for Vessels and Teams

The **Marine Coordination Center (MCC)** is a central component of offshore wind projects, overseeing the coordination and safety of all maritime activities. In its role as a communication point between vessels, offshore teams, and other project participants, the MCC ensures that all work can be conducted efficiently and safely. It acts as an interface for information exchange and plays a crucial role in resource management and risk minimization.

11.6.1 Tasks and Responsibilities of the MCC

The MCC is responsible for planning, monitoring, and coordinating all vessel movements within the project area. This includes transport and installation vessels, Crew Transfer Vessels (CTV), and Service Operation Vessels (SOV). It ensures that all ships perform their tasks in line with the project schedule and that all maritime activities are coordinated to avoid conflicts and delays.

A central task of the MCC is monitoring vessels' position data using real-time tracking systems. This allows the MCC to view the current position of all ships in the project area at any time and ensure they are moving within designated zones. Simultaneously, the MCC ensures compliance with safety guidelines, such as safe distances from underwater structures, other ships, or sensitive environmental areas.

11.6.2 Communication with Vessels and Offshore Teams

The MCC functions as a communication hub for all ships and offshore teams. It ensures that important information—such as weather warnings, navigation instructions, or operational changes—is quickly and precisely communicated to affected parties. This communication is especially important to minimize delays or safety risks due to unexpected events.

An example of the MCC's role is coordinating crew transfers. The MCC ensures that crews are brought to platforms or wind turbines at scheduled times and that return transports occur on time. This involves considering the availability of vessels, weather conditions, and safety regulations.

11.6.3 Coordination of Maritime Safety Measures

The safety of all maritime activities is the MCC's top priority. It continuously monitors weather conditions to recognize potential hazards like high waves, strong winds, or poor visibility early. In dangerous conditions, the MCC informs affected ships and teams and coordinates the interruption or adjustment of activities if necessary.

Additionally, the MCC is responsible for coordinating emergencies. In the event of an accident or hazard, such as a collision or medical emergency, the MCC organizes appropriate rescue measures and ensures clear communication among involved parties. This includes coordination with external rescue services or authorities.

11.6.4 Integration into Project Management

The MCC works closely with other project participants, such as commissioning management and the Operation Control Center (OCC). It ensures that maritime activities are seamlessly integrated into the overarching project plan and that no conflicts arise with other work, such as cable or turbine installation. This close collaboration is crucial for effective resource utilization and ensuring project progress.

Another aspect of integration is documentation. The MCC maintains detailed logs of all vessel movements, work, and safety incidents. This documentation is important not only for traceability and analysis but also for compliance with legal requirements and audits.

11.7 Differentiation Between OCC and MCC

The **Operation Control Center (OCC)** and the **Marine Coordination Center (MCC)** have different but complementary roles in the communication of offshore wind projects.

- **OCC**: Focuses on the control and operational aspects of the installations, including coordination of switching operations, monitoring electrical systems, and support with technical issues. It is responsible for communication between onshore and offshore teams, as well as for the safety and tracking of all work on electrical and mechanical systems.
- **MCC**: Responsible for maritime coordination, monitoring all vessel movements, crew transfers, and transport operations in the project area. It primarily communicates with vessels and offshore teams, ensures compliance with maritime safety guidelines, and coordinates rescue measures in emergencies.

Both centers jointly contribute to project safety by overseeing and organizing the technical processes (OCC) and the maritime infrastructure (MCC).

Part V
Economic Aspects of Commissioning Management

The economic aspects of an offshore wind project are just as critical as the technical challenges. Alongside the technical and operational project management, a Financial Project Manager is indispensable, bearing the main responsibility for the overall project budget. Close and seamless collaboration between the technical project management and the finance department is of paramount importance. Only through this cooperation can financial risks be identified, mitigated, and controlled while simultaneously ensuring the project's economic efficiency.

Part V highlights the key elements of successful economic management, including:

Risk Assessment and Mitigation: Evaluating potential financial risks and developing strategies to minimize their impact on the project.

Budgeting and Resource Planning: Establishing a comprehensive budget and effectively allocating resources to meet project objectives.

Contractual Challenges: Navigating legal complexities, ensuring compliance with contractual obligations, and managing claims.

The success of a project significantly depends on how well these aspects are integrated and jointly managed. By fostering collaboration between technical and financial teams, projects can achieve their goals more efficiently and sustainably.

Chapter 12
Economic Risk Assessment

Economic risk assessment is a central component in the planning and execution of an offshore wind project. It forms the basis for informed decision-making and enables the minimization of potential financial losses through preventive measures. This chapter highlights the importance of collaboration with the Risk Manager, examines the different types of risks and threats, and explains how risks are represented in the project budget.

12.1 Collaboration with the Risk Manager

A successful approach to economic risk assessment requires close collaboration between the Commissioning Management and the Risk Manager. While the Commissioning Management identifies technical and operational risks, the Risk Manager is responsible for translating these risks into economic impacts and developing appropriate strategies to control them.

Effective information exchange is essential in this context. For example, the Commissioning Management provides data on potential delays, material failures, or unexpected technical challenges. The Risk Manager evaluates these scenarios and prepares corresponding risk analyses that consider both probabilities and potential financial impacts.

Joint workshops and regular coordination between both parties help develop a holistic view of the risks. This collaboration ensures that technical and economic risks are not considered in isolation but are evaluated in the overall context of the project.

C. Hoppe, *Commissioning Management in Offshore Wind Projects*,
SpringerBriefs in Applied Sciences and Technology,
https://doi.org/10.1007/978-3-032-05358-9_12

12.2 Risks and Threats

In the context of offshore wind projects, **Risks** and **Threats** are two closely related but distinct concepts, each with specific impacts on the project. Risks refer to potential uncertainties that can have both positive and negative effects on the project. They encompass events or conditions that can bring either additional benefits (opportunities) or harm (losses). Threats, on the other hand, are specific risks with clearly negative impacts that can jeopardize the project, such as safety–critical incidents, financial losses, or delays.

12.2.1 Positive and Negative Impacts of Risks

Risks in offshore wind projects are not inherently negative. When properly managed, they can also present opportunities. An example of a positive impact is optimizing construction time due to unexpectedly favorable weather conditions, allowing work to be completed faster than planned. Such a development could lead to savings in personnel and vessel resources.

An example of a negative impact would be the occurrence of extreme weather conditions leading to installation delays. This could result in additional charter costs for vessels and extended deployment times for offshore personnel, significantly increasing project costs.

12.2.2 Risks to Be Evaluated in Offshore Wind Projects

Offshore wind projects are subject to a variety of risks due to their complexity and long duration. The following examples illustrate the types of risks that can generally occur, especially during the commissioning phase:

- **Weather Risks**: Offshore environments are heavily influenced by unpredictable weather conditions. Strong winds, high waves, or storms can hinder access to installations and cause delays. These risks are particularly high during winter months.
- **Temporal Risks (Delays, Catch-up, and Congestion)**: Delays in the delivery of components, installation, or completion of preliminary work (e.g., cable laying) can lead to "congestion," where different teams or tasks must be carried out simultaneously, increasing complexity and the risk of errors.
- **Clash Situations**: These arise when multiple activities are conducted in the same area or on the same installation. An example would be performing cable tests and mechanical work on a turbine simultaneously, which can lead to safety and coordination issues.

- **Personnel Risks**: This includes the availability of qualified personnel, the health and safety of employees, and the effects of high workloads, especially during critical project phases.
- **Legal and Regulatory Changes**: Due to long project durations, legal frameworks can change, affecting project planning and implementation. An example is the introduction of new environmental regulations requiring additional testing or project modifications.
- **Technological Risks**: New technologies or insufficiently tested components could lead to unexpected problems that impair installation or operation.

This list is not exhaustive but serves as an orientation for the types of risks typically encountered in offshore wind projects. Each project must conduct an individual risk analysis to identify specific challenges.

12.2.3 *Risk Mitigation*

Risk mitigation is a central component of risk management and includes strategies to reduce the likelihood and impact of risks. Key measures include:

- **Proactive Planning**: Early planning and scenario analyses help recognize potential risks and develop measures before they occur. An example is adjusting the construction schedule to seasonal weather conditions to avoid delays due to storms or rough seas.
- **Contractual Safeguards**: Clear contractual agreements can transfer risks, such as delays or additional costs, partially to suppliers or subcontractors.
- **Schedule Flexibility**: Incorporating buffer times into the project schedule allows for the absorption of delays without jeopardizing the entire timeline.
- **Escalation Management**: Establishing clearly defined escalation processes ensures that risks are quickly identified and effectively addressed.
- **Training and Education**: Regular training of employees, especially in safety and technical areas, reduces the risk of errors and accidents.
- **Monitoring and Reporting**: Continuous monitoring and regular reports help track risks and assess their development. Modern tools, such as risk registers or digital dashboards, can support this.

Distinguishing between risks and threats and systematically assessing their impacts is essential for the success of an offshore wind project. Risks are unavoidable but also offer opportunities when managed proactively. By identifying specific risks, applying risk mitigation measures, and closely collaborating with the Risk Manager, Commissioning Management can contribute to minimizing financial losses and securing project success. The combination of precise planning, flexibility, and continuous monitoring forms the basis for effective risk management in offshore wind projects.

12.3 Representation of Risks in the Budget

Presenting risks in the project budget is a crucial step to account for their potential financial impacts and ensure the project's economic stability. This is typically achieved by establishing a risk budget, also known as a contingency budget, which is set up and managed by the Risk Manager.

A risk budget is based on a detailed analysis of identified risks and their probabilities. Amounts are calculated based on the expected costs and the likelihood of a risk occurring and are added to the overall project budget. This budget serves as a buffer to cover unforeseen costs without endangering the project schedule or other financial commitments.

Integrating risks into the budget can take various forms:

- **Direct Risk Positions**: Specific budget items are set up for identified risks, such as bad weather periods or technical challenges.
- **General Risk Contingency**: A total amount is reserved for unforeseen risks that may arise during the project duration.
- **Specific Provisions**: These provisions relate to known risks with a high probability of occurrence, such as additional charter times for vessels during bad weather periods.

In addition to financially planning for risks, regular monitoring of the risk budget is necessary. This is done by comparing planned and actual costs, with deviations documented and analyzed. The results feed back into the risk analysis to adjust the budget if necessary.

By systematically integrating risk assessment into budget planning, projects can better prepare for uncertainties and maintain financial control. This proactive approach allows for timely adjustments and ensures that the project remains economically viable even when faced with unforeseen challenges.

Chapter 13
Budgeting and Resource Planning

Budgeting and resource planning are central elements of a successful offshore wind project. They form the basis for the effective utilization of personnel, materials, and financial resources to achieve project goals on time and within the prescribed budget. In addition to financial aspects, organizational, logistical, and sustainability considerations also play a decisive role. This chapter highlights the importance of sustainable planning, dealing with contingencies, the fundamentals of personnel and resource planning, and additional cost factors such as training, tools, and personal protective equipment (PPE).

13.1 Sustainability in Project Planning

Sustainability in offshore wind projects is not only an environmental goal but also a key to efficient and economical planning. In terms of personnel deployment, this means optimizing travel routes and minimizing travel times. By employing rotating teams that regularly alternate, not only is employee workload reduced, but continuous availability of skilled personnel is also ensured. Where possible, local labor can be integrated to further reduce travel times and costs.

The sustainable use of material resources is another important aspect. Reusable materials and precise demand planning help avoid waste and overstocking. For example, modular offshore components can be returned and used for other projects after installation is complete. Environmental regulations, such as the storage of hazardous substances or the transport of sensitive materials, must be considered in planning to meet legal requirements and minimize environmental risks.

C. Hoppe, *Commissioning Management in Offshore Wind Projects*,
SpringerBriefs in Applied Sciences and Technology,
https://doi.org/10.1007/978-3-032-05358-9_13

Digital tools for optimizing logistics and material transport offer further opportunities to promote sustainability and efficiency. Real-time monitoring of material movements, combined transports, and proactive warehousing not only minimize environmental impacts but also project costs. Thus, sustainability in planning represents a win–win situation for both the environment and cost-effectiveness.

13.1.1 Including Contingencies in Planning

Contingencies are an unavoidable part of offshore wind projects, as work is heavily dependent on external factors such as weather and logistics. Precise planning requires the inclusion of risks and buffer times to cushion delays or unforeseen costs. Bad weather periods, especially in autumn and winter months, pose one of the greatest challenges. Time buffers for delays or unforeseen logistical bottlenecks are essential to keep the schedule on track.

In addition to time buffers, redundancy in materials and personnel should also be planned. Having spare parts available for critical components like transformers or cable connections can reduce downtime. Standby teams that can be mobilized at short notice are also an important part of flexible and resilient resource planning.

The budget for unforeseen costs, often referred to as a risk budget or contingency budget, typically ranges from 10 to 15 percent of the total budget. This budget serves to cover additional costs due to re-deliveries, vessel delays, or other unforeseen events. Regular monitoring and adjustment of the risk budget can ensure that contingencies do not throw the project off financial balance.

13.1.2 Planning Fundamentals for Personnel and Resource Planning

Personnel and resource planning in offshore wind projects is particularly complex, as it must consider a multitude of factors. A key aspect is the consideration of transport times and waiting times. The distance between the port and the project site affects not only logistics but also the deployment time of employees. In typical offshore projects, these distances often range between 30 and 50 km, which can mean transit times of several hours per day, depending on weather conditions and the type of vessels used.

Additionally, weather windows must be planned to minimize delays due to rough seas or strong winds. This is especially true for critical phases such as turbine installation or commissioning of electrical systems. Hands-on-tool times, that is, the effective working time on site, are another important factor. These times include the actual duration of work minus setup and dismantling times, safety briefings, and transport times. Precise planning that considers these factors is crucial to maximize work efficiency.

The use of digital tools to monitor and optimize resource utilization can bring significant advantages here. Real-time data on work progress, availability of personnel and materials, and weather forecasts help to make planning flexible and efficient. Multi-purpose vessels that can transport both personnel and materials are another approach to reduce waiting times and costs.

13.1.3 Training Costs, Tools, and Personal Protective Equipment

In addition to direct costs for personnel and materials, training, tools, and personal protective equipment (PPE) also play a significant role in resource planning. Employees must be regularly trained to ensure they meet the technical and safety requirements for offshore work. These trainings cover topics such as occupational safety, high-voltage systems, first aid, and emergency procedures. The costs for these trainings can be considerable, especially when specialized certifications are required.

Tools and PPE also represent a not insignificant cost factor. High-quality and regularly maintained tools are necessary to carry out work efficiently and safely. PPE such as helmets, safety glasses, harnesses, and offshore survival suits must meet the highest standards to ensure employee safety. Since this equipment is subject to high wear, it must be regularly inspected and replaced if necessary, which must also be accounted for in resource planning.

13.1.4 Consideration of National Working Time Regulations

Another decisive factor in personnel planning is national and regional working time regulations. Offshore work often falls under specific legal provisions that prescribe maximum deployment times and minimum rest periods. The 14/14 model, where employees are on duty for 14 days and then have 14 days off, has proven effective in many projects. It offers a good balance between working time and rest but may not be feasible in all countries or projects.

Compliance with these working time regulations affects the planning of replacement crews and the organization of shift changes. In regions with strict working time requirements, additional teams must be planned to ensure legally mandated rest periods. Flexible shift models adapted to the specific requirements of the project and region are indispensable here.

By comprehensive budgeting and resource planning that considers sustainability, contingencies, training, and legal requirements, offshore wind projects can be implemented efficiently and safely. This planning lays the foundation for a smooth project execution and ensures that all resources are optimally utilized to achieve project goals.

13.2 Deployment Planning of Offshore Teams

Deployment planning of teams in offshore wind projects is a dynamic and central component of commissioning management. It is primarily based on the commissioning list, which sets out all planned work and their chronological sequence. Effective deployment planning combines short-term adjustments with long-term strategies and requires consideration of numerous factors such as progress reports, resource availability, and project specifications. The close integration of planning elements such as Daily Progress Reports (DPRs), 3-Day Look-Aheads, mid-term and long-term planning is crucial to implement the project efficiently and cost-effectively.

13.2.1 The Commissioning List as the Basis of Planning

The commissioning list is the central document reflecting the current status and planned work in the project. It contains detailed information on individual tasks, including their prioritization, dependencies, and estimated duration. This list serves as the starting point for deployment planning and forms the basis for all subsequent planning stages. As the list is constantly updated through progress reports and feedback from teams, it remains a living and dynamic tool.

13.2.2 Short-Term Planning: Daily Progress Reports and 3-Day Look-Aheads

Short-term deployment planning relies heavily on Daily Progress Reports (DPRs) and 3-Day Look-Ahead reports. DPRs provide a current overview of work progress and identify potential problems or delays. Based on this information, deployment planning can be adjusted daily to respond to unforeseen events such as weather changes or material shortages.

The 3-Day Look-Aheads, on the other hand, offer a preview of the coming days and help plan resources efficiently. These reports consider task dependencies, availability of personnel and materials, and possible weather windows. By combining DPRs and 3-Day Look-Aheads, a flexible planning tool is created that allows short-term adjustments while ensuring a certain level of stability.

13.2.3 Mid-term Planning: Mid-term Resource Planning

Mid-term planning, often referred to as Mid-Term Resource Planning, covers a period of about two to eight weeks and serves the overarching coordination of resources. In this phase, deployment plans are coordinated with other project areas to avoid conflicts or bottlenecks. This particularly concerns the availability of vessels, heavy equipment, and key personnel.

A central aspect of mid-term planning is considering project progress compared to the original schedule. Scenarios such as P50 (base scenario) and P90 (worst-case scenario) are used here to assess the probabilities of delays or additional requirements. These scenarios help to plan realistic buffer times and resource capacities to minimize risks while keeping costs in mind.

13.2.4 Long-Term Planning: Long-Term Resource Planning

Long-Term Resource Planning extends over several months and includes the strategic planning of the entire resource needs for the project duration. This phase is particularly important to secure long-term contracts for personnel, vessels, or equipment. At the same time, basic deployment plans are developed here, which serve as orientation for mid-term and short-term planning.

A central focus is on the optimal utilization of resources. Long-term planning must ensure that capacities are neither over- nor under-utilized to control costs and maximize efficiency. Again, P50 and P90 scenarios play an important role in realistically assessing the impacts of delays, additional requirements, or unexpected risks.

13.2.5 Interaction of All Factors

The different planning phases—short-term, mid-term, and long-term—are closely interconnected. Successful deployment planning requires continuous information flow between the levels so that current developments can be integrated into the plans. Particularly, the regular updating of the commissioning list through DPRs and 3-Day Look-Aheads is crucial to keep the planning process flexible and dynamic.

Planning must also be coordinated with external factors such as weather conditions, resource availability, and legal requirements. Close collaboration among project participants is essential to avoid overlaps or conflicts. This also includes communication with the Marine Coordination Center (MCC), the Operation Control Center (OCC), and on-site teams.

13.2.6 Cost-Time-Benefit Trade-Off

In deployment planning, the trade-off between cost, time, and benefit is always at the forefront. While very flexible planning can cause additional costs, overly rigid planning can lead to delays and inefficiencies. The key lies in balance: resources must be deployed to achieve the maximum possible efficiency without excessively burdening costs.

The approach to deployment planning is often specified by project management. Some projects place more emphasis on cost-optimized planning, while others focus on maximum flexibility and risk mitigation. It is the responsibility of commissioning management to operationalize these specifications while ensuring effective implementation.

Deployment planning in commissioning management is a complex, iterative process influenced by a multitude of factors. Close coordination between planning levels, the use of scenarios like P50 and P90, and the continuous updating of the commissioning list are crucial to implement projects successfully and efficiently. A well-thought-out interplay of all factors enables optimal resource utilization while keeping costs and risks under control.

Chapter 14
Contract and Claim Management

Contract and claim management is an essential component of any offshore wind project, serving to clarify contractual issues and address potential conflicts between parties at an early stage. This chapter provides an overview of some fundamental aspects of this subject but does not claim legal validity. It is important to emphasize that I am not a lawyer and therefore cannot offer legal advice. For binding information and decisions, specialized attorneys or experienced contract law experts should be consulted.

In well-structured projects, an independent contract and claim management department is indispensable. This department works closely with the technical departments to clarify contractual questions, report hindrances, and review the responsibilities of the involved parties. Collaboration between these departments and project management is crucial to minimize legal risks and ensure successful project execution.

14.1 Transfer of Risk

The term **"transfer of risk"** describes the point in time when the risk of loss or damage to a work passes from the contractor to the client. According to the FIDIC contract conditions, this transition typically occurs upon acceptance of the work. Until that point, the contractor bears both the risk of damages and the responsibility for complete performance.

In certain cases, however, the transfer of risk can occur before formal acceptance, such as when the client puts the work into use or unjustifiably refuses acceptance. From the time of risk transfer, the client is liable for accidental events that could damage or destroy the work, and the contractor is entitled to the agreed remuneration. A clearly defined and documented transfer of risk is crucial to avoid later misunderstandings or conflicts.

C. Hoppe, *Commissioning Management in Offshore Wind Projects*,
SpringerBriefs in Applied Sciences and Technology,
https://doi.org/10.1007/978-3-032-05358-9_14

14.2 Notice of Hindrance

A **notice of hindrance** is a central mechanism by which the contractor informs the client of circumstances that could impair or delay the proper fulfillment of contractual services. Under the FIDIC contract conditions, the contractor is obliged to issue this notice promptly and in writing. It should include an exact description of the hindering circumstances, the time the hindrance began, and an assessment of the anticipated duration.

The timely and precise issuance of a notice of hindrance is essential to be able to claim extensions of construction time or additional compensation. If the contractor fails to issue this notice, claims for extension or additional remuneration may be lost unless the hindrance was obvious to the client. It is therefore essential that the contractor, despite the hindrance, takes all reasonable measures to continue the work and promptly informs the client when the work is resumed.

14.3 Causer and Responsible Party

A central aspect of contract and claim management is determining which party is responsible for a hindrance or damage. The **causer** is the party whose action or omission triggered the hindrance or damage, while the **responsible party** is the one liable for the legal and financial consequences. Depending on the contractual conditions, these two roles can be identical or different.

Determining responsibility requires a detailed analysis of the facts, including consideration of contractual agreements and applicable legal provisions. The responsible party may be obliged to pay damages, while the non-responsible party may be entitled to an extension of construction time or additional remuneration. It is important that the party making claims bears the burden of proof for the cause of the hindrance. Clear documentation of all processes is therefore indispensable.

A well-founded contract and claim management requires not only legal expertise but also close collaboration with the technical and operational departments of a project. Through clear processes, detailed documentation, and close coordination among the involved parties, risks can be minimized and legal conflicts avoided. Projects with a well-equipped contract and claim management department are capable of proactively and professionally handling these complex issues, bringing them to a successful conclusion.

Part VI
Industry-Wide Standards and the Future

The offshore wind industry stands at a pivotal point. Beyond the implementation of industry-wide standards, the future will be shaped by technological innovations and sustainable strategies. This part of the book examines the regulatory requirements and standards that serve as the foundation for safe and efficient offshore wind projects and explores the trends and visions that will influence the industry in the coming decades.

Not only is the importance of automation and digitalization highlighted, but there is also an increasing focus on sustainability and reducing the ecological footprint. Innovations such as floating foundations, higher voltage levels in intra-park layouts, and the black start capability of wind farms will decisively shape the future. Additionally, the concept of energy islands as a possible solution for integrating large offshore wind capacities is examined.

This part illustrates how commissioning management can benefit from these developments and what challenges need to be addressed to shape a sustainable and resilient energy future.

Chapter 15
Regulatory Requirements and Standards

Regulatory requirements and standards form the foundation for the planning, construction, and commissioning of offshore wind projects. They ensure that these projects can be carried out safely, efficiently, and sustainably. At the same time, they are highly complex and vary depending on the region, legislation, and the stipulations of stakeholders such as Transmission System Operators (TSOs), regional authorities, and legislators. This chapter provides an overview of some essential norms and standards but does not claim to be exhaustive. It is crucial to engage thoroughly with the specific requirements of each project to avoid unpleasant surprises.

15.1 ISO and IEC Standards for Offshore Wind Projects

International standards, particularly those of the International Organization for Standardization (ISO) and the International Electrotechnical Commission (IEC), form the basis for the technical and operational safety of offshore wind projects. These standards are only examples and must always be reviewed in the context of the project and the respective region.

- **ISO 29400:2020, Ships and marine technology—Offshore wind energy—Port and marine operations**, describes requirements and guidelines for the planning and execution of port and marine operations. The focus is on the safe and efficient handling of materials and personnel in these critical areas.
- **IEC 61400-3-1:2019, Wind energy generation systems—Design requirements for offshore wind turbines**, sets the minimum requirements for the design of fixed offshore wind turbines. The goal is to ensure the structural integrity and long-term operational capability of the installations.

C. Hoppe, *Commissioning Management in Offshore Wind Projects*,
SpringerBriefs in Applied Sciences and Technology,
https://doi.org/10.1007/978-3-032-05358-9_15

- **IEC TS 61400-3-2:2019, Design requirements for floating offshore wind turbines**, complements the previous standard with specific requirements for floating wind turbines. This standard considers additional challenges such as movement caused by wind and waves.

These examples illustrate how specifically international standards address the particular requirements of offshore wind projects. Nevertheless, it is important to examine the full range of relevant standards to ensure that no requirements are overlooked.

15.2 DNV as an Independent Certifier with Recognized Standards

The Norwegian organization **DNV (Det Norske Veritas)** is a globally recognized independent certifier that develops specific standards for offshore wind projects. These standards cover all project phases and serve as guidance for developers, operators, and authorities.

- **DNV-ST-N001, Marine Operations and Marine Warranty**, offers detailed guidelines for the planning and execution of transport, installation, and decommissioning of offshore wind farms. These standards are especially important for the safety and efficiency of complex maritime operations.
- **DNV-SE-0190, Project Certification of Wind Power Plants**, specifies the requirements for project certification of offshore wind turbines. The flexible approach of this standard allows for consideration of project-specific peculiarities.
- **DNV-SE-0073, Project Certification of Wind Farms According to IEC 61400-22**, provides a structured certification scheme that ensures offshore wind farms meet the requirements of IEC standards. It is an important foundation for contractual security and the approval of such projects.

DNV standards are widely used and offer clear guidelines that facilitate compliance with international norms. They ensure that projects meet the highest safety and quality standards.

15.3 Requirements from Transmission System Operators (TSOs) and Regional Authorities

A crucial aspect of offshore wind projects is the specific requirements from TSOs and regional authorities. These vary greatly depending on the region and stakeholders, making a precise analysis of applicable norms and stipulations indispensable.

- **Grid Connection Conditions and Grid Compliance Study**

TSOs set technical requirements for grid connection, including voltage levels, frequency stability, and reactive power compensation. To ensure that an offshore wind farm meets these requirements, a so-called Grid Compliance Study is conducted. This includes load flow analyses, short-circuit current analyses, and dynamic simulations, among others. The results of this study are essential to guarantee grid stability and to obtain the final grid connection approval from the TSO.

- **Approval Procedures by Regional Authorities**

Regional authorities impose requirements ranging from environmental impact assessments to technical verifications. Potential impacts on the environment, maritime infrastructure, and adjacent industries are examined. These procedures are often lengthy and require precise preparation.

- **Operational and Maintenance Requirements**

TSOs and regional authorities may set requirements for the operation and maintenance of offshore wind farms to ensure grid stability and supply security. This includes stipulations for regular inspections, fault reports, and system availability.

The regulatory requirements and standards in offshore wind projects are complex and multifaceted. In addition to international norms like ISO and IEC, the stipulations of independent certifiers like DNV and the requirements of TSOs and regional authorities play a decisive role. A detailed engagement with these standards is indispensable for the successful implementation of a project. Only through careful planning and adherence to all stipulations can it be ensured that offshore wind projects run smoothly and can be operated successfully in the long term.

Chapter 16
The Future of Commissioning Management

The offshore wind energy industry will undergo profound changes in the coming years, shaped by technological innovations, increasing performance demands, and a growing awareness of sustainability. Commissioning management, as a critical phase in the commissioning of offshore wind farms, will need to adapt to these developments to ensure efficiency, safety, and environmental compatibility.

16.1 Automation and Digitalization

In the future, automation and digitalization will revolutionize commissioning management. **Artificial Intelligence (AI)** will play a central role by enabling real-time monitoring of installations, supporting predictive maintenance strategies, and optimizing operational processes. By employing machine learning, anomalies can be detected early, increasing the availability of assets and minimizing downtime. For example, AI will be capable of analyzing data from various sensors and identifying patterns that indicate potential problems [1].

Despite these technological advances, the human factor will remain indispensable. The commissioning manager will continue to contribute expertise, experience, and decision-making abilities that cannot be replaced by technology alone. The future will lie in the symbiotic collaboration between humans and machines, with digital tools serving as support to enhance efficiency and safety. It is expected that the commissioning manager will be able to better visualize complex systems and conduct training more effectively through the use of **Augmented Reality (AR)** and **Virtual Reality (VR)**.

C. Hoppe, *Commissioning Management in Offshore Wind Projects*,
SpringerBriefs in Applied Sciences and Technology,
https://doi.org/10.1007/978-3-032-05358-9_16

16.2 Sustainability and Reduction of Ecological Footprint

Sustainability will remain a central concern in the offshore wind industry. Studies have shown that offshore wind farms save up to **50 times more CO_2** over their lifetime than is emitted during their manufacturing, construction, operation, and decommissioning [2].

Nevertheless, challenges will persist. The expansion of offshore wind energy could have significant impacts on marine flora and fauna. It will therefore be essential to carefully consider environmental aspects and take measures that minimize the ecological footprint. For example, innovative foundations could be developed that require less invasive installation methods, thereby reducing impacts on the seabed.

16.3 Trends and Innovations in the Offshore Wind Industry

Offshore wind energy will be characterized by groundbreaking innovations in the coming years. **Floating foundations** will enable the installation of wind turbines in deeper waters and farther from the coast (Fig. 16.1). This allows for the exploitation of new wind resources that have remained unused until now. This technology will play a significant role especially in regions like the Mediterranean or off the coasts of Japan and the USA, where deep waters predominate. With increasing commercialization, it is expected that floating wind farms will make up a substantial share of global offshore wind capacity.

A particularly exciting technological approach is the **Self-Sustainable Turbine (SST)** by Siemens Gamesa Renewable Energy (SGRE). This concept aims to develop wind turbines capable of supplying themselves with energy in idle mode without relying on external power sources. This capability is especially relevant during the construction and commissioning phases when the grid is not yet available, as well as in maintenance scenarios to ensure the operation of critical systems such as lighting, control, and communication. While the technology is still in an early development stage, it could significantly reduce operational costs in the future and simplify logistics for temporary power supply systems.

The challenges in implementing SSTs currently lie in scalability and integration into existing system designs. It is expected that in the coming years, more prototypes will be tested to demonstrate the functionality and reliability of this technology under real conditions. If the concept proves successful, it could sustainably change the way wind farms are operated and maintained.

Fig. 16.1 Floating foundations, https://infrastructureusa.org/floating-foundations-a-game-changer-for-offshore-wind-power/, Nov. 2024

16.3.1 Increasing Voltage Levels in the Inter-array Layout

With the continuous increase in the output of modern wind turbines, which now exceed the **15-megawatt** mark, it will be inevitable that voltage levels of at least **110 kV** will become standard within wind farm cabling. The use of these higher voltage levels will be necessary to handle the increased energy flow efficiently and to minimize transmission losses. However, this imposes significant demands on the infrastructure, including new cable technologies, more powerful transformers, and enhanced protection systems.

The integration of such systems will require close coordination between developers, manufacturers, and grid operators. It is foreseeable that the higher complexity of the inter-array layout will also necessitate new standards and certifications to ensure the safety and reliability of these systems.

16.3.2 Outlook

Through innovations such as self-sustaining turbines in the absence of grid supply, floating foundations, and the introduction of higher voltage levels, the offshore wind industry will enter a new era. These technologies promise to increase efficiency, reduce operational costs, and enhance the flexibility of wind farms. Nevertheless, the industry will continue to face challenges that must be addressed in the coming years with a strong focus on research, development, and collaboration.

16.3.3 *Black-Start Capable Wind Farms in the Commissioning Context*

Black-start capability will play a key role for offshore wind farms in the future, particularly in the context of the resilience of modern power grids. A black start describes the ability of a facility to restart after a complete power outage without external energy sources. In commissioning management, this presents a special challenge, as this capability must be tested and ensured at an early stage.

16.3.4 *Relevance for Commissioning*

For commissioning, black-start capability means that not only the regular commissioning must be tested but also special scenarios where the facility operates autonomously. This includes testing the interaction between turbines, energy storage systems, and control units in **island mode**. The goal is to ensure that the systems are capable of establishing stable grid frequency and voltage, even without external energy sources.

16.3.5 *Technological Requirements and Testing*

During the commissioning process, the following tests are the focus:

- **Energy Storage Systems**: These are tested for their ability to ensure a constant power supply during a black start. Storage solutions like batteries or flywheels must be operated under realistic conditions.
- **Control Systems**: The software and hardware of control units must be capable of precisely regulating energy production. Here, it is particularly tested whether the systems can function autonomously without external signals.
- **Island Mode Operation**: Island mode operation, where the wind farm operates independently of the grid, is tested for stability. Protection mechanisms and control strategies must ensure that there are no unexpected voltage or frequency deviations.

16.3.6 Future Perspective

In future offshore wind farms, black-start capability will become a standard test in commissioning. It not only enables higher grid stability but also strengthens the role of wind farms as active elements of energy supply. For commissioning management, this means expanded responsibility and greater integration into strategic grid planning.

More will be written on this topic in the next chapter.

References

1. S.J. Omexom, *Damit sich was dreht: Wie KI die Zuverlässigkeit der Offshore-Windkraftanlagen verbessert* (2023). Von https://www.energie.de/et/news-detailansicht/nsctrl/detail/News/damit-sich-was-dreht-wie-ki-die-zuverlaessigkeit-der-offshore-windkraftanlagen-verbessert abgerufen
2. Oersted, *Wie groß ist der CO_2-Fußabdruck der Offshore-Windenergie?* (2024). Von https://orsted.de/gruene-energie/offshore-windenergie/fakten/co2-fussabdruck abgerufen

Chapter 17
Black-Start Capable Wind Farms: The Current State

Currently, only a few offshore wind farms possess the capability for a black start. The focus so far has been more on integrating large energy storage systems that serve as buffers for grid stability. Projects like the **Kriegers Flak Combined Grid Solution** in the Baltic Sea demonstrate initial approaches to enhancing grid stability through offshore wind farms.

Further progress is being made by combining wind farms with hydrogen electrolyzers. These systems could not only store excess electricity but also be used for the black-start process. However, the industry is still in its early stages, and many challenges—especially in control technology and energy storage—need to be addressed.

17.1 The Future of Black-Start Capable Offshore Wind Farms

In the coming years, technological innovations are expected to significantly improve the black-start capability of offshore wind farms. The following developments could form the foundation for this advancement:

- **Advancements in Energy Storage**: New technologies like solid-state batteries, flywheels, or high-performance capacitors will increase the reliability and capacity of energy storage systems.
- **Improved Control Systems**: Progress in artificial intelligence and machine learning will enable better anticipation of complex grid requirements and adjustment of turbine control accordingly.
- **Grid-Independent Start Sequences**: Through autonomous control strategies, wind farms could establish grid frequency without external assistance and support other energy sources during reactivation.

C. Hoppe, *Commissioning Management in Offshore Wind Projects*,
SpringerBriefs in Applied Sciences and Technology,
https://doi.org/10.1007/978-3-032-05358-9_17

17.1.1 Challenges and Potential

Implementing black-start capability in offshore wind farms will involve significant costs, particularly for the necessary storage solutions and adaptation of existing infrastructure. Regulatory requirements also pose an additional hurdle, as many national grid operators have specific provisions for black-start capable facilities.

Despite these challenges, black-start capability offers enormous potential for offshore wind farms. It could strengthen the resilience of power grids and further solidify the role of offshore wind energy as a central component of the energy transition. Additionally, it would facilitate the integration of renewable energies into existing grids and reduce dependency on fossil backup systems.

17.2 The Energy Island

TenneT, the Dutch-German transmission system operator, has developed the concept of an **"Energy Island"** in the North Sea to promote the expansion of offshore wind energy and strengthen Europe's energy supply. This visionary project aims to construct an artificial island as a central hub for offshore wind farms, efficiently feeding the generated electricity into European power grids.

17.2.1 Concept of the Energy Island

The planned Energy Island is intended to be built on the **Dogger Bank**, a shallow sandbank approximately 125 km off the east coast of Great Britain. Covering an area of around six square kilometers, it would serve as a central platform for connecting multiple offshore wind farms. The island would be equipped with converter stations that transform the alternating current generated by the wind farms into direct current, minimizing losses during long-distance transmission [1].

17.2.2 Advantages of the Project

- **Efficient Power Transmission**: By pooling the generated electricity on the island and subsequently converting it to direct current, transmission losses can be reduced, and efficiency increased.
- **Cost Savings**: The central infrastructure allows multiple wind farms to connect jointly to the power grid, reducing costs for cables and converter stations.

- **International Collaboration**: The island could serve as a hub for the power grids of several European countries, including the Netherlands, Germany, Denmark, and Great Britain, thereby improving cross-border energy supply [2].

17.2.3 Current Status and Outlook

TenneT, together with partners like **Energinet**, **Gasunie**, and the **Port of Rotterdam**, has formed the consortium **"North Sea Wind Power Hub"** to investigate and advance the project's feasibility. The goal is to achieve a capacity of **30 gigawatts** of offshore wind energy by 2030 and provide up to **180 gigawatts** in the long term [3].

Although the project is still in the planning phase, it shows the potential to significantly transform offshore wind energy in Europe and make a substantial contribution to achieving climate goals.

References

1. D. Neumann, *Größter Windpark der Welt: Energie-Insel soll 30 Gigawatt Ökostrom erzeugen* (2018). Von https://www.futurezone.de/science/article213025381/groesster-windpark-der-welt-energie-insel-soll-30-gigawatt-oekostrom-erzeugen.html abgerufen
2. S. Hanke, *Cobra, übernehmen Sie!* (2017). Von https://energiewinde.orsted.de/energiewirtschaft/energie-insel-strom-verteilkreuz-doggerbank-cobracable abgerufen
3. N. Weinhold, *Künstliche Inseln als Wasserstoffspeicher für Offshore* (2019). Von https://www.erneuerbareenergien.de/technologie/onshore-wind/tennet-konsortium-kuenstliche-inseln-als-wasserstoffspeicher-fuer-offshore abgerufen

Conclusion on Commissioning Management

In reflecting upon the journey of commissioning management, it becomes evident that experience alone is insufficient without its diligent application. As Werner von Siemens aptly stated, "We do not lack experience, but rather the adherence to it." This emphasizes the importance of integrating lessons learned into every facet of our work. By consistently applying past insights, we not only avoid repeating mistakes but also foster innovation and efficiency. Embracing this philosophy enables commissioning managers to navigate complexities more effectively and contribute to the continuous advancement of the offshore wind industry.

Commissioning management is a discipline that continually evolves, and its significance grows with each new offshore wind project. This book provides a comprehensive foundation to understand the central aspects and challenges of this complex task. However, it does not claim to cover all facets exhaustively. Rather, it is intended as a guide and incentive to delve deeper into the diverse topics of commissioning management and to develop your own approaches.

Over the past ten years, I have personally developed more than 60 specific documents for commissioning management and countless other commissioning documents. Presenting all of them here would not only exceed the scope but also influence the creative approach of each manager. Commissioning management thrives on flexibility and the ability to adapt documents to the specific requirements of a project. No two projects are alike, and a rigid approach would be neither effective nor sustainable.

The strength of commissioning management lies in its adaptability. Documents and processes must be designed to meet the requirements of each individual project without compromising efficiency or safety. Successful commissioning management therefore requires not only technical knowledge but also leadership qualities, strong communication skills, and the ability to work in multidisciplinary teams.

In this book, we have illuminated the most important subject areas: from the fundamentals and processes to the tools and practical implementation, as well as economic and forward-looking aspects. We have tried to maintain a balance between

C. Hoppe, *Commissioning Management in Offshore Wind Projects*,
SpringerBriefs in Applied Sciences and Technology,
https://doi.org/10.1007/978-3-032-05358-9

technical depth and strategic overview. The examples and analyses presented are meant to inspire, not to be dogmatic. They are intended to help you find and adjust your own approach.

The future of commissioning management will be characterized by innovations, sustainability, and increasing complexity. Technologies like artificial intelligence, automation, and new energy storage solutions will further change the field and open up new opportunities. But despite all technological advancements, humans remain the heart of every project. The success of a commissioning manager will continue to depend on their ability to act flexibly, solve problems, and inspire teams.

I hope that this book not only provides you with valuable insights and tools but also motivates you to actively shape the future of commissioning management. May your projects be successful, your teams work harmoniously together, and your visions become reality. Commissioning management is a demanding but also incredibly fulfilling task—and you are an important part of this development.

Glossary

Glossary

3-Day Look-Ahead Report (3DLAR) A planning tool that provides a detailed preview of the upcoming three days in a project, focusing on scheduled tasks, required resources, and potential challenges to ensure efficient short-term planning.

Authorized Person (AP) An individual who has been granted the authority to carry out specific tasks, such as switching operations in electrical systems, under the supervision of a Senior Authorized Person (SAP).

Augmented Reality (AR) A technology that overlays digital information onto the real world, enhancing the user's perception and interaction with their environment, often used for training and visualization in commissioning processes.

Black-Start Capability The ability of a power-generating facility, such as an offshore wind farm, to restart and restore operations independently without relying on external power sources after a total shutdown.

Commissioning List A central document detailing all planned commissioning activities, including tasks, priorities, dependencies, and estimated durations, serving as the basis for deployment planning.

Commissioning Management The process of planning, executing, and overseeing the commissioning phase of a project to ensure systems and components meet operational requirements and function as intended.

Contingency Budget A portion of the project budget reserved to cover unforeseen costs or risks that may arise during the project lifecycle, typically ranging from 10 to 15% of the total budget.

Daily Calls Short, daily meetings (usually 15-30 min) focused on discussing the previous day's progress, current challenges, and plans for the day to keep the project team aligned.

C. Hoppe, *Commissioning Management in Offshore Wind Projects*,
SpringerBriefs in Applied Sciences and Technology,
https://doi.org/10.1007/978-3-032-05358-9

Daily Progress Report (DPR) A report documenting the daily progress of a project, including completed tasks, ongoing activities, challenges faced, and plans for the next day.

Det Norske Veritas (DNV) An independent Norwegian certification body providing globally recognized standards and certifications, particularly for offshore wind projects.

Energy Island A proposed artificial island concept serving as a central hub for connecting multiple offshore wind farms, facilitating efficient power transmission to onshore grids.

Escalation Management A structured process for addressing and resolving issues or risks that cannot be managed at the operational level, involving higher levels of management or specialized personnel.

Floating Foundations Innovative support structures for wind turbines that allow installation in deep waters where traditional fixed foundations are not feasible.

Grid Compliance Study An analysis conducted to ensure that a power-generating facility meets the technical requirements set by Transmission System Operators (TSOs) for grid connection, including voltage levels and frequency stability.

Hands-On-Tool Time The actual effective working time that personnel spend on-site performing tasks, excluding transit, setup, and briefing times.

Marine Coordination Center (MCC) An organizational unit responsible for planning, monitoring, and coordinating all maritime activities in an offshore project, including vessel movements and crew transfers.

Minutes of Meeting (MoM) A documented record of discussions, decisions, and action items from a meeting, distributed to participants and relevant stakeholders for reference and accountability.

Operation Control Center (OCC) A central hub in offshore projects that monitors and controls all electrical and mechanical systems, coordinates switching operations, and serves as a communication point between onshore and offshore teams.

P50 and P90 Scenarios Statistical measures used in project planning:

- **P50 Scenario**: Base case where there is a 50% probability that the estimated value (like cost or time) will not be exceeded.
- **P90 Scenario**: Conservative case with a 90% probability that the estimated value will not be exceeded, accounting for higher uncertainty or risk.

Personal Protective Equipment (PPE) Safety equipment worn by personnel to minimize exposure to workplace hazards, including helmets, safety glasses, harnesses, and protective clothing.

Risk Budget See **Contingency Budget**.

Risk Mitigation Strategies and actions taken to reduce the likelihood and impact of identified risks on a project.

Risk Register A tool used to document all identified risks, including their probabilities, potential impacts, and mitigation measures.

Senior Authorized Person (SAP) A highly qualified individual with the authority to oversee and approve critical tasks such as switching operations in electrical systems.

Service Operation Vessel (SOV) A specialized ship used in offshore wind projects to provide accommodation, transfer personnel, and support maintenance activities at sea.

Switching Order A formal document or instruction detailing the steps and procedures for performing electrical switching operations to ensure safety and compliance.

Task Force Meeting A focused meeting involving specialists assembled to address and resolve specific issues or challenges within a project.

Transmission System Operator (TSO) An entity responsible for the transmission of electricity through high-voltage power lines, ensuring grid stability and managing grid connections for power-generating facilities.

Weekly Progress Report (WPR) A comprehensive report summarizing the project's progress over a week, including milestones achieved, risks identified, resource utilization, and strategic planning for the upcoming week.

Wind Turbine Generator (WTG) The complete system of a wind turbine, including the blades, generator, tower, and associated equipment, used to convert wind energy into electrical power.

Uncited Reference

1. Internationales Wirtschaftsforum Regenerative Energien (IWR)/IWR.de GmbH, Premiere Offshore Windenergie auf der Zielgeraden: Frankreich baut ersten Offshore Windpark im Mittelmeer (2024). Von https://www.offshore-windindustrie.de/news/ticker/premiere-offshore-windenergie-auf-der-zielgeraden-frankreich-baut-ersten-offshore-windpark-im-mittelmeer-artikel7090 abgerufen

C. Hoppe, *Commissioning Management in Offshore Wind Projects*,
SpringerBriefs in Applied Sciences and Technology,
https://doi.org/10.1007/978-3-032-05358-9

Index

C. Hoppe, *Commissioning Management in Offshore Wind Projects*, SpringerBriefs in Applied Sciences and Technology,
https://doi.org/10.1007/978-3-032-05358-9

Made in the USA
Monee, IL
13 September 2026

62975227R00090